Finite and Infinite Mathematics

Sets, Numbers, Lines, Equations, Probability

Finite and Infinite Mathematics
Sets, Numbers, Lines, Equations, Probability

by Dr. Sanford Aranoff

My goal and desire in teaching is to convey the beauty, logic, and understanding of the mathematical concepts to the students. I taught a course in Finite Mathematics for several years, and have seen where students get confused with concepts. Throughout the semesters, I have emailed students summaries and ideas. I have collected these, with the idea of my being able to communicate with more students than I am able to in a classroom. In addition, I hope that my future students will benefit from this book. Read this slowly. Go one step at a time. You may jump around sections, but read each section trying to understand the ideas.

Hope this book helps you with your understanding of mathematics and other subjects!

Good luck in your studies!

-Dr. S. Aranoff

Preface

After using this book during the 2011 – 2013 semesters at Rider University, we found necessary changes and typos. This version includes these changes.

Table of Contents

1. Introduction

This chapter discusses the author's background, and goals of the book.

This material is
Copyright © 2013 by Dr. Sanford Aranoff.

The current revision was written on 07/11/2013. This contains minor corrections.

In addition to Microsoft Word, this was created with Help & Manual, which automatically creates page numbers, table of contents, index, and an Adobe pdf file. MathType was used to create mathematical expressions.

1.1 Introduction

For a number of years I have been teaching a first-year mathematics course *Finite Mathematics* at Rider University, NJ, using the textbook *Mathematical Ideas* by Miller, Heeren, and Hornsby. Although I feel this is an excellent text, there have been many ideas and examples that I had to clarify to the students. In addition to explaining material in class, I also wrote these additions and emailed them to the students. My goal as a teacher is to help students to be able to understand the material so that they can be able to learn new things on their own.

What I am now doing by writing is explaining my ideas to an audience wider than the classroom. My goal is to help future students who use Miller's text.

The chapters in this book more or less match the chapters in Miller's book. I cannot be precise, as they rearrange the chapters when revising the book, which they do every few years.

1.2   My background

My Ph.D. is in theoretical physics from New York University. Theoretical physics is basically mathematical physics. The stress is understanding the universe based upon developing and using principles and comparing the results with empirical evidence. I spent a number of years as a physics professor at Rutgers, NJ. After this, I went into software development with the emphasis on defense, as an extension of my mathematics and rational training and experience. Recently I returned to academia, as an Adjunct Associate Professor at Rider University, NJ.

My career choice of academia is based upon my belief that the university gives people the opportunity to use their rational thinking to the maximum along with the ability to verify the results of these thoughts. The goal is not to develop a better product that will make money, or to develop a better weapon to help defend our country, but to understand the world in which we live, and to pass this understanding on to others.

My goal is to teach the beauties of 20th century and 21st century science and mathematics to people, to help other people to better understand things, especially mathematics and science. On December, 2010, Rider University gave me an award for "your service and dedication to the freshman class at Rider."

1.3   Success is valid thinking

An organism is successful to the extent to which its basic needs are
met. People need more than basic physical needs. We need to know
how to think clearly and rationally, and to properly communicate these
thoughts. To the extent that we are clear and rational, we are
successful in this human endeavor. Success is valid thinking and
proper communication and appreciation of ideas. Failure is thinking
that violates rules of correct thinking, or ignoring empirical evidence
contradicting thoughts.

Examples of such failures are most medieval religious thought, such as
Catholic or Islamic thinking. Examples of successes are 19th century
development of electromagnetism and thermodynamics, and 20th
century development of quantum mechanics.

"There was never naught,

"There was always thought." –Robert Frost

1.3.1   Why study mathematics?

A student commented about mathematics. He asked why we need to explain things, instead of just doing problems and getting the results. Indeed, this is how high schools teach mathematics. Students learn a bunch of unrelated tricks and pass tests showing that they know how to use the tricks. My answer is that mathematics is rational thinking, a skill that everyone needs in order to survive, both as individuals and as a nation. Here is why we need rational thinking.

Several years ago, I taught a course that involved mortgages and interest payments. To make the discussion simple, I am going to use easy numbers. Let us say a person takes a $100,000 mortgage for 10 years. The bank gives the person the money, and the person then can buy a house. The person has to pay more than $10,000 a year, for there are interest payments. Let us say the interest is 5%. This means that each year, in addition to the $10,000, he has to pay $500. Part of this money is used to pay the bank for the expenses of the mortgage. There is another reason for the interest payments.

When one applies for a mortgage, he has to supply financial details, such as his employment history. The bank makes a determination of the person's ability to make the mortgage payments. Banks have experience over a century of the statistics of income and ability to pay. The bank may decline the applicant for fear that this person may default. In addition, under the best of circumstances, some people default. For example, the bank may know that out of 100 people 1 person will default. This means that the other 99 people have to pay enough interest to cover the $100,000 loss due to default.

Along came the federal government under President Clinton who asked the banks to give mortgages to people they would otherwise not give them, so that more people can become homeowners. Clinton used the CRA, the Community Reinvestment Act. The banks went along with the government. The reasoning was that the more mortgages they sold, the more money the banks would make. Unfortunately, they made an error with their mathematics. They forgot the losses they would make that would not be covered by the insurance payments.

I told my students in my class several years ago that this is simply irrational. It is impossible for the banks to continue doing this. I said to the students they must understand mathematics, and not be stupid.

Well, as we know, the banks indeed lost all their money and cried to the government for bailouts.

This shows the need to understand mathematics and to be rational.

There is another aspect to the story. Many individuals signed up for mortgages, even though their income did not justify it. They probably said that they do not know mathematics, and trusted the banks that know mathematics. Big mistake! They lost their homes and everything, when they could have lived comfortably in rented apartments.

Each of us must be responsible for ourselves.

A few years ago on vacation, I saw an osprey nest. One of the parents would be in the next while the other went fishing, and they took turns. Here is what I said to my students. Imagine I am the daddy bird. I am teaching the chicks to fly. (I wave my hands). I then sternly say to them that when they fly in the air, they are on their own, with no one to help them. They will have to get their own food and seek their own shelter.

We study mathematics in order to better be rational thinkers, to help us survive in the world where people want to take what we have and expect us to give it without saying anything. Let us not fall for smooth talk, but think carefully.

1.3.2 Mixed successes

Contrasting with these successes, Einstein's theories of special and general relativity are a mixed bag of both successes and failures. Consider special relativity, which Einstein published in 1905. This is very easy to understand if explained properly. Unfortunately, it is often not explained properly, because Einstein himself did not fully understand it.

In the 19th century, Maxwell developed electromagnetism, basing it upon four postulates. He showed that light is an electromagnetic wave, with the speed of light being defined from the constants of electromagnetism. Einstein developed his special relativity theory postulating (making an arbitrary statement serving as a foundation of the theory) that the speed of light is constant in all inertial systems (coordinate systems moving at constant velocities). This leads to the surprising conclusion that the speed of light measured by equipment on the ground and equipment in a moving truck should be the same. If we throw a baseball and measure its speed, equipment on a moving truck will measure a different speed than equipment on the ground. The fact that the speed of light is different from the speed of baseballs surprised many people.

Unfortunately, Einstein erred in saying "the constancy of the speed of light is a postulate." The error is that this statement is *not* arbitrary, but a *conclusion* of Maxwell's theory. We can derive special relativity by postulating that all physical theories must be the same no matter in which inertial system we view them. It is unnecessary to postulate the constancy of the speed of light.

Allow me to repeat and summarize this confusing point. Maxwell derived the speed of light from other equations. This means that the *speed of light is a theorem*, similar to theorems in geometry which we prove using the postulates. Einstein's error was saying the *speed of light is a postulate.* **Mathematics students must know what the postulates are and what the theorems are, and not confuse theorems with postulates.** This book helps clarify this distinction.

Using special relativity, we can derive Maxwell's results with only one postulate, instead of four. For this, Einstein deserves the greatest praise. This is a scientific success. However, Einstein's error caused considerable unnecessary confusion on a subject that is intrinsically simple.

What surprised me was that as late as 1970, decades after Einstein's publication of special relativity, I had to publish papers showing errors physicists had in their understandings. This was, in my opinion, a clear failure of the university community to apply the rules of logical thinking concerning special relativity. One reason is the failure of Einstein and others to properly explain the logical structure of this theory of physics. In my opinion, this failure is because Einstein had a full time research position with no teaching responsibilities. Had he stood in front of a small class of young college students, he would have been forced to think clearly how to develop the theory so that the young people would understand the ideas. Instead, Einstein went on lectures, explaining to large groups or to other physicists, without the feedback of classroom interactions and by looking at homework and tests. The scientist or mathematician must teach undergraduates in order to fully and properly understand the material, IMHO.

One of the reasons for this decades-long misunderstanding is the awe many people had to Albert Einstein. Emotional feelings like awe have no place in rational thinking in mathematics and science. Students have feelings of awe to their school mathematics teachers, and this makes it difficult for many to look beyond this and try to understand what the basic principles are and what the necessary conclusions are.

The purpose of this book is to encourage correct rational thinking about mathematics, and to avoid pitfalls such as occurred with this example from physics.

1.4　References of my papers

"Postulates of Special Relativity," Sanford Aranoff, *Physics Essays* (a peer-reviewed journal published through the American Institute of Physics), **26**, 1, March (2013).

"The Aura of Einstein and General Relativity," *philica.com*, article number 192, Aranoff, S. (2010).

"Basic Assumptions and Black Holes," Sanford Aranoff, *Physics Essays*, **22**, 559 (2009).

"Professors Substituting In High School", S. Aranoff, *Gifted Education Press Quarterly*, Winter (2013).

"Lessons We Can Learn from Bill Gates' Dropping Out of Harvard University",
S. Aranoff, *Gifted Education Press Quarterly*, Fall (2011).

"How is a Teacher of the Gifted Supposed to Teach?"
S. Aranoff, *Gifted Education Press Quarterly*, Fall (2009).

"To Educate the Gifted, We Need to Stress Basic Principles,"
S. Aranoff, *Gifted Education Press Quarterly*, Winter (2009).

"What Young Teachers of the Gifted Need to Know and Do,"
S. Aranoff, *Gifted Education Press Quarterly*, Winter (2008).

"Equilibrium in Special Relativity," Sanford Aranoff,
IL Nuovo Cimento, **10B,** 155-171 (1972).

For other papers and work discussing contemporary errors in scientific thinking, please see my website:
http://www.analysis-knowledge.com/msgTeaching.htm.

1.4.1   Books that I wrote

One book that I wrote is *Teaching and Helping Students Think and Do Better.* This is based upon my experiences observing teaching, and my experiences teaching and tutoring.

Another book is *Rational Thinking, Government Policies, Science, and Living.* Rational thinking starts with clearly stated principles, continues with logical deductions, and then examines empirical evidence to possibly modify the principles. Mathematics is rational, as the proofs must be logical. We need to be rational in order to best deal with the problems and issues we face in life.

1.5 Some quotes

Here are some quotes that I have collected over the years.

"We need to talk and act as if we truly believe that the humanities, the arts, mathematics, and science matter and underlie the deepest foundations of a democratic society. The ultimate foundation of any society ought to be the human imagination, honed to the greatest degree and in the company of its faithful companion, curiosity."
– D. M. Randel, *Dædalus*, Winter (2009).

"Mathematics is human imagination taken to its limits."

"Critical thinking is the Socratic habit of articulating questions and gathering relevant information in order to make reasonable judgments. This is why I say that if a student fails to ask questions during class, the student may likely fail the course. Each student must make judgments about the material."

"If we do not challenge the gifted to become the best that they can be, we are losing out as a society.

"The future of any society depends on the character of its elites. America's future depends upon how we educate the academically gifted. Our elites are smart but not wise. They may mean well, but do not know virtue. "
– Charles Murray, *Real Education: Four Simple Truths for Bringing America's Schools Back to Reality*

"Solving problems isn't just a mechanical procedure; it calls for more than a set of skills. It requires attitudes and dispositions - the courage needed to acknowledge the existence of a problem that has to be dealt with; the patience and persistence required when a problem isn't easily resolved; willingness to risk, to seek help and to give it, to accept personal responsibility, and to admit error."

– Donald Arnstine, quoted in *Why is Corporate America Bashing our Public Schools?*, K. Emery and S. Ohanian (2004).

Vincent C. LoPresti in *Science*, 12/13/1987, p. 730,
Letter to the Editors:
"Quality science teaching at middle and high school levels requires a deep conceptual understanding of the themes that pervade natural science, ...integrating conceptual themes, ...to produce a society of informed citizens capable of some understanding of emergent technologies."

Clayton C. Denman. "An appalling number of young people enter college with little understanding of both social and technological aspects of world."

"If a nation expects to be ignorant and free
It expects what never was and never will be." – Thomas Jefferson

"The trouble with the world is that the stupid are cocksure and the intelligent are full of doubt." – Bertrand Russell

"There's a tendency on the part of the public to see science as something very remote because it's so arcane and difficult to understand. But the uses of science are everywhere and it's very important for all of us to recognize that and to strengthen our institutions that relate science to society."
– David A. Hamburg, President, Carnegie Corp. of New York (1987).
To this statement about science I would add mathematics – SA.

"Train for the known, educate for the unknown." -Seth K. Powell
Students are often being trained in mathematics , but being educated in mathematics . We need creative thinking skills to solve the unknown problems that we will be faced with in the future.

1.6 Symbols

Here are symbols we use in math:

∃   **There exists.**

∍   **Such that.** In English, a is divisible by b if there is a k such that $a \div b = k$.

      In math, a is divisible by b if $\exists k \ni \dfrac{a}{b} = k$, where $a, b,$ and k are natural numbers.

∈   Element of a set.

      E.g., 1 is an element of the set $\{1, 2\}$.

      In math: $1 \in \{1, 2\}$.

∀   **All.** For example, in English:

      the **inverse property** is that

      for every real number not zero

      (that is, for *all* numbers) there exists a

      reciprocal that is also a real number

      such that the product of the number

      and the reciprocal is one.

      In math: if R is the set of real numbers,

      $\forall a \neq 0,\ a \in R,\ \exists \dfrac{1}{a} \in R \ni a \times \dfrac{1}{a} = 1.$

Δ   **Change.** E.g., $\Delta y = y_2 - y_1.$

(a,b) **Ordered pair.** For example,

the coordinates of a point (x, y).

$\aleph_0$ **Aleph null**, the cardinality of the set of natural

numbers, also called countable infinity.

E.g., $n(\{1, 2, 3, \ldots\}) = \aleph_0$.

Let us not confuse this with ∞.

$\aleph_0$ is the last integer, as we count $1, 2, 3, \ldots$

c The continuous infinity of real numbers. The

number of points on a line between 0 and 1 is c.

iff :: **if and only if.** *A iff B* means *A* implies *B*

AND *B* implies *A*. $A \leftrightarrow B$.

For example: $\dfrac{a}{b} = \dfrac{c}{d}$ iff $a \times d = c \times b$.

$\therefore$ **Therefore.**

$f(x)$ **Function.** f is a function of x.

Given x, this gives f.

E.g., $f(x) = mx + b$.

We may write it as $y(x)$.

!   **Factorial.** $n! = n(n-1)(n-2)\ldots 1$

This is true for $n > 0$. For $n = 0$, $0! = 1$.

1.6.1 Summation and products

Here are definitions of the math symbols.

Σ for summation of terms

Π for multiplication of factors

These are capital Greek letters.

Mathematicians use capital Sigma to simplify how expressions look.
Suppose we have an expression like $1 + 2 + 3 + 4 + 5$.

Instead of writing 5 numbers, we can write one number preceeded by the Sigma symbol

$$\sum_{i=1}^{5} i = 1 + 2 + 3 + 4 + 5.$$

This involves less symbols to write. Likewise,

$$\sum_{i=1}^{5} x_i = x_1 + x_2 + x_3 + x_4 + x_5.$$

It also focuses our attention on the key idea, the thing that is changing.

$\sum_{i=0}^{n} f(i)$ means we sum the function $f(i)$ from $i = 0$ to $i = n$.

Written out, this is $f(0) + f(1) + f(2) + \cdots + f(n)$.

E.g., $\sum_{i=1}^{3} (i+1) = 2 + 3 + 4.$

1.6.2 Sets

Although sets are discussed below in Chapter 3, set symbols are given here so that you will be able to find all the symbols in one place. Here sets are written in capital letters, such as A, and elements of sets in lower case, such as a.

{ } Set notation. For example, $A = \{1, 2, 3\}$.

$n(A)$ The **cardinality** of the set A. For $A = \{1, 2, 3\}$, $n(A) = 3$.

~ **Equivalence.** Sets A and B are equivalent:

$A \sim B$ if $n(A) = n(B)$

$\emptyset$ The **empty** set. $\emptyset = \{ \ \}$.

$\cup \cap$ **Union** and **intersection**

Union means *or*, intersection means *and*.

$A \cup B$ means the set of elements either in A *or* in B.

$A \cap B$ means the set of elements in both A *and* B.

– **Difference.**

$A - B$ means the set of elements in A *and not* in B.

$\{1, 2, 4\} - \{2, 3\} = \{1, 4\}$.

$\subseteq$ **Subset.** For example, let A and B be sets.

$A \subseteq B$ if $\forall a \in A$ then $a \in B$.

Another definition of equality:

$A = B$ if $A \subseteq B$ and $B \subseteq A$.

$\subset$ **Proper subset.** If $A \subset B$, then $A \subseteq B$ and $A \neq B$.

' **Complement.** A' is the set of elements in the universe U *and not* in A.

/ Slash through symbol means not. Examples:

$\neq$ means not equal, or $\not\subseteq$ means not a subset.

$A \not\subseteq B$ means that A is not a subset of B.

$a \notin A$ means a is not an element of the set B.

1.6.3   Greek

Here are the Greek letters. You are now in college, and need to know Greek. Mathematicians like to use Greek letters for symbols. You can see these letters in programs like Microsoft Word.

$$A \; B \; \Gamma \; \Delta \; E \; Z \; H \; \Theta \; I \; K \; \Lambda \; M \; N \; \Xi \; O \; \Pi \; P \; \Sigma \; \; T \; Y \; \Phi \; X \; \Psi \; \Omega$$

$$\alpha \; \beta \; \gamma \; \delta \; \epsilon \; \zeta \; \eta \; \theta \; \iota \; \kappa \; \lambda \; \mu \; \nu \; \xi \; o \; \pi \; \rho \; \varsigma \; \sigma \; \tau \; \upsilon \; \varphi \; \chi \; \psi \; \omega$$

Note that there are two ways to write lower case sigma.

Here are the names and sounds. Alpha, beta, gamma, delta, ... omega.

A	α	alpha	a	"father"
B	β	beta	b	
Γ	γ	gamma	g	
Δ	δ	delta	d	
E	ε	epsilon	e	"end"
Z	ζ	zêta	z	
H	η	êta	ê	"hey"
Θ	θ	thêta	th	"thick"
I	ι	iota	i	"it"
K	κ	kappa	k	
Λ	λ	lamda	l	
M	μ	mu	m	
N	ν	nu	n	
Ξ	ξ	xi	ks	"box"
O	o	omikron	o	"off"
Π	π	pi	p	
P	ρ	rho	r	
Σ	σ, ς	sigma	s	"say"
T	τ	tau	t	
Y	υ	upsilon	u	"put"
Φ	φ	phi	f	
X	χ	chi	ch	"Bach"
Ψ	ψ	psi	ps	
Ω	ω	omega	ô	"grow"

1.7   Gödel's Incompleteness Theorem

Gödel showed that any system within which arithmetic can be developed, is essentially incomplete. In other words, given any consistent set of arithmetical axioms, there are true mathematical statements that cannot be derived from the set. This is analogous to the set of natural numbers. This set is incomplete, for no matter how large the set, we can add 1 to the largest number to get a new number.

A mathematical system consists of postulates and the logical conclusions. A scientific theory is a mathematical system along with empirical verifications. We can imagine all known knowledge like this:

Gödel showed that there are true statements we cannot derive from the known postulates. That is, there is an infinite number of such statements. Known knowledge is a subset in the infinite sea of true statements. The picture of reality actually looks like this:

We deal with this reality beyond what is known by music, emotions, and such, for we cannot use logic. We must deal with the known reality the best we can to optimize our lives and society.

2. Problem solving

When students think about math, they think of problem solving. They think that if they can solve problems, they are good at math. While it is true that problem solving is an important part of mathematics, it is not all math.

2.1 What is mathematics?

Before discussing problems, let us discuss what mathematics actually is.

What is mathematics? Students invariably respond saying that mathematics is numbers. This is not correct. A mathematical system is a collection of *arbitrary self-consistent statements*. A student enters class and sits down. Why did she choose this seat and not that seat? Her choice is arbitrary, for she can sit in any empty seat. The arbitrary part of mathematics makes it fun, and the consistency requirement makes it challenging.

The demanding logic and effort of learning mathematics helps all of us to think clearly and rationally, critically important to our lives as adults and as a society. The challenges facing all of us in the future are huge. We need to think rationally to deal with these challenges. The study of mathematics will help us develop and use tools for the required rational thinking we will face in our future lives.

Mathematics is like a diamond – very hard, but each way you look at it you see something different.

Mathematics is very beautiful, and the correct understanding gives great pleasure, just as does good music or fine art. In addition, there are many applications of mathematics. We will discuss the beauty of mathematics and the power of the applications.

Music is like mathematics in the following ways:

- Both have to follow rules.

- Both are truth - music is enjoyment, and math is consistency. Neither is scientific truth.

- Mathematicians like to generalize and be creative to get new mathematical concepts. Musicians like to be creative and write new music.

- Mathematics communicates ideas that cannot be written in words. Music expresses things that cannot be expressed in words.

- Mathematicians enjoy the math for its beauty, and wish to share the beauty with others. Artists and musicians enjoy the beauty of their creations and wish to share it with others.

- There are many emotions musicians have when performing. There are only two emotions mathematicians have when doing mathematics. One is the unhappy bored feeling when the mathematician knows he is wrong. The other is the happy, elated emotion when correctly understanding a new idea or creating a new concept. We need to pay attention to our emotions. If we do not feel good and happy, we may be wrong, and should spend extra time thinking about what we were doing.

2.1.1 Postulates and axioms

A postulate is often incorrectly defined as "Something taken to be true without proof"

(*American Heritage Talking Dictionary*. Copyright © 1997 The Learning Company, Inc.).

A postulate is also defined as "A proposition regarded as self-evidently true without proof,"
Mathworld, http://mathworld.wolfram.com/Axiom.html.

The word "axiom" is a slightly archaic synonym for postulate.

The correct definition is, in my opinion:
A postulate is an arbitrary statement, consistent with the other postulates of the system.
Arbitrary means you can say whatever you please.
A mathematical system is a collection of arbitrary (made-up) statements that do not contradict each other.

Of course, the statements in different mathematical systems do not have to agree. We ask only that a given mathematical system be consistent.

Here is the dictionary definition of *postulate*:

1. Something assumed without proof as being self-evident or generally accepted, especially when used as a basis for an argument: "the postulate that there is little moral difference between the superpowers"
(Henry A. Kissinger).
2. A fundamental element; a basic principle.
3. Mathematics - an axiom.
4. A requirement; a prerequisite.

What are the basic principles of mathematics? Answer: arbitrary statements, i.e., postulates.

Here is the dictionary definition of *axiom*:

1. A self-evident or universally recognized truth; a maxim: "It is an economic axiom as old as the hills that goods and services can be paid for only with goods and services"(Albert Jay Nock).
2. An established rule, principle, or law.
3. A self-evident principle or one that is accepted as true without proof as the basis for argument; a postulate.

We see that axiom is the same as postulate. The definition of postulate is axiom, and the definition of axiom is postulate. This looks a bit circular to me!

A student wrote, "A postulate is a math assumption that can be proven." This is not correct. We use postulates to prove other things. We do not prove postulates.

2.1.2   What mathematics is

I ask students, "What is mathematics?" The answer is frequently, "Numbers." No! Instead, mathematics is logic. Let us examine geometry for example.

Geometry is based upon postulates.

Mathematics is a human activity. All that we ask from a mathematical system is consistency.

Euclidean geometry is a consistent system. A modification of Euclid's fifth postulate will also generate a consistent system. This leads to non-Euclidean geometry. An example of this is *hyperbolic geometry*. Given the postulates, we apply logic and prove theorems. If we can prove two theorems that contradict, then we have shown that the postulates are not consistent, and so that mathematical system is not valid.

High school geometry teachers teach Euclidean geometry as if this were the only possible consistent mathematical system. Although non-Euclidean geometry is too much for high school students, they should be briefly exposed to the ideas. This exposure will help them properly understand the true subtle meanings of mathematics.

In addition to the logical need for this understanding, the real world is non-Euclidean. This is the case near very massive objects with huge gravitational fields. The earth's gravitational field is small enough so that the Euclidean approximation is valid for all our purposes.

2.1.3   Example of what mathematics is

Here is an example of applying the definition of mathematics, as arbitrary statements.
To solve the equation $x + 3 = 5$, we move the 3 to the right-hand side, and get $x = 2$.

This we learned in middle school. Let us understand the words *"solve"* and *"equation"*. An equation is a statement with an equal sign. This idea is very simple, and yet students are confused about it. Let us look at a statement without the equal sign
such as $x = 3$.
This is an *expression*. The word "solve" means to
find the value of the variable x in an equation.

When we say "move the 3 to the right-hand side", there is a question.

How do we move a number, something that exists only in our minds? I can move the chalk from one board to another, but how do I move a number? Numbers are not part of reality. The answer is that we need to use our axioms and postulates. (These two words mean the same thing).

Here are a few postulates:

- Any expression or number equals itself.
- Given an equation, we can multiply each term of the equation by a number or expression and get a valid equation.
- Given two valid equations, we can add them to get a third valid equation.

Here is how we apply the postulates:

$x + 3 = 5$ Given equation

$-3 = -3$ Equality postulate

$x = 2$ Add two equations and get another equation: postulate

In summary, when we say "move the 3 to the right-hand side", we are actually performing the above three steps. We are doing a mental shortcut.

2.1.4   Abstractions

One very important aspect of mathematics is abstraction. The school stress on rote, problem solving, and how to do things hides the beauty and necessity of mathematical abstraction. Mathematicians like to look at things and try to abstract the basic ideas.   Abstraction encourages generalizations and new understandings. Finally, it helps thinking, as we start with the very simple and progress to the complicated, instead of starting with the complicated.

Let us give an example with the idea of a circle. Students know what a circle is, by taking a compass, opening it up to the desired radius, sticking the sharp point on the paper, and drawing the circle. Oh, what a pretty circle! Draw more circles, and see what pretty designs we can get!

What are the essential features of our circle? One is the sharp point. We need to specify the location of this point. We draw a coordinate system on the paper, and specify the x, y coordinates of the point. The next is the orientation of the paper, whether flat on the desk, or vertically if we drew the circle on the blackboard. Mathematically, we need to specify the plane. Finally, we need to give a number, the number of inches the radius. In short, we need a plane, a coordinate system with a specified origin, the coordinates of the point, and the number of inches in the radius.

If we are discussing 2-dimensional geometry, then the orientation of the plane is assumed. We also usually assume the location of the coordinate system. In this case, a circle is simply a point and a number.

Note what we have done. Instead of thinking of this round thing that we drew, we think of a point and a number. In a similar fashion, we know that given two points there is one and only one line going through these points. This is discussed below. We can abstract a line to two points. Giving two points is the same as giving the line.

Consider this generalization. A problem could be given two points find the equation of the line, giving things like the slope. Next, let one point be the center of a circle, with the distance between the points the radius. Find the equation of the circle. This should be very easy. However, students get confused, as they do not see the relation with points defining a line and a circle. They fail to appreciate the abstraction.

2.1.5   How to bake a cake

This site discusses how to bake a cake.
http://www.wikihow.com/Bake-a-Cake.

Go to this website and look at it. Note that there are so many, many steps! Some women who bake great cakes cannot tell others how what they did. This difficulty in communication is because they look at each step as a separate concrete item, and there are too many items to memorize. However, if we abstract the steps, and allow one to question a step for greater details, then it is easy. You will have to learn how to abstract concepts. Mathematics is abstraction.

Here is how I abstract the cake recipe:
1. Choose the type of cake to bake.
2. Get the ingredients.
3. Mix, put in pan in the oven
4. Take out of oven

Four steps! Easy to remember! Very different from the twenty or so steps a woman is likely to say!

Each step can be broken down into smaller steps. For example, step 3 really is:
- mixing dry first
- then wet
- Put flour in pan
- preheat oven
- etc.

This is how we must think about mathematics. Try to understand it from the highest level of abstraction. Then look at each level and go into greater detail. This book is written in the fashion. The highest level of abstraction is the chapters. Each chapter is divided into subheadings. Some subheadings are further subdivided.

2.2 Math anxiety

Here are some discussions in the literature on the topic, journal articles and Internet sites.

2.2.1 Journal articles

Journal of Educational Psychology, **82**, No. 1, 60-70 (1990).
The point here is "students' performance expectancies predicted subsequent math grades." If you think you will not do well in the mathematics class, you likely will not do well. This article makes a point that I feel is obvious. The solution is just as obvious. Do not think you will not do well. How do you change your thinking? What I do is to tell myself what to think. Tell yourself that you will do well in the mathematics course.

Here is another article.
Journal of Counseling Psychology, **25**(5), 441-448. doi: 10.1037/0022-0167.25.5.441 (1978).
Results indicate that math anxiety occurs frequently among college students and that it is more likely to occur among women than among men and among students with inadequate high school math backgrounds. Higher levels of math anxiety were related to lower mathematics achievement test scores, higher levels of test anxiety, and higher levels of trait anxiety. Implications for the identification and treatment of math-anxious students and for the process of educational/vocational counseling are discussed here.

2.2.2   How to fix anxiety

This is an interesting article that gives an excellent discussion of the problem, along with recommendations to students. This was written by B. Sidney Smith.

http://www.mathacademy.com/pr/minitext/anxiety/

Here are excerpts:

- When I look at a math problem, my mind goes completely blank. I feel stupid, and I can't remember how to do even the simplest things.

- I've hated math ever since I was nine years old, when my father grounded me for a week because I couldn't learn my multiplication tables.

- In math there's always one right answer, and if you can't find it you've failed. That makes me crazy.

- Math exams terrify me.

- I've never been successful in any math class I've ever taken. I never understand what the teacher is saying, so my mind just wanders.

- Some people can do math – not me!

Here is how I would respond to the above points.

My solution to the mind going blank is to remember to start from the beginning, and to realize that you may not see the end. When we hike a trail, we know where to begin, and often do not see the end.

Hating math because of the inability to learn multiplication tables is unfortunate, for math is not multiplication tables, but logical thinking starting from principles. I find the need to reiterate this to students.

In math there are often many right answers. Sometimes a student will see an answer that the professor did not see. However, students frequently feel that I want only one right answer.

A person terrified by exams needs to know how to take exams. This is discussed below. Look away for a moment, think of nothing, and then look back at the material. One should study this way also.

A student who does not understand the teacher, and lets his mind wander, is not doing his job as a student. Your job is to understand the material, and this requires you to raise your hand and ask questions. If you fail to ask questions in class you may fail the course.

Finally, never say you cannot do math. Say that you will try to understand the beginnings and know how to proceed.

2.2.3 Myths about math

Here are some more comments from the math academy site by Smith, with my paraphrasing and additions.

For nearly seventy years, teaching methods have relied on a behaviorist model of learning, a paradigm which emphasizes learning-by-rote; that is, memorization and repetition. In mathematics, this meant that a particular type of problem was presented, together with a technique of solution, and these were practiced until sufficiently mastered. The student was then hustled along to the next type of problem, with its technique of solution, and so on. The ideas and concepts which lay behind these techniques were treated as a sideshow, or most often omitted altogether.

Some say that the aptitude for math is inborn. Actually, all of us are capable of reasoning with abstract ideas. Fear of getting "the wrong answer" interfers with our freedom to think and to reason.

Mathematics is a science of ideas, not an exercise in calculation. You can be good at mathematics while poor at calculation.

Some say that mathematics requires logic, not creativity. This is not so. The great mathematicians, indeed, are poets in their soul.

Many students focus on getting the right answer. They err in thinking this is what the professor wants. This is a big mistake. Focus on understanding.

A psychologist made an experiment with a monkey. He put a banana in a cage attached to the ceiling with a string. He figured that there were three ways the monkey may try to get the banana. When he started the experiment, the monkey started screaming, and the psychologist went over to see what was going on. The monkey jumped on the psychologist's shoulders and got the banana. The psychologist did not think of this way! You see, the teacher may not know all the answers!

2.2.4   Taking possession

Here are final comments from the math academy site by Smith.

The first step, and the one without which no further progress is possible, is to recognize that math anxiety is an emotional response.

Begin by understanding that your feelings of math anxiety are not uncommon, and that they definitely do not indicate that there is anything wrong with you or inferior about your ability to learn math.

Don't wait until after the lecture – raise your hand or your voice the minute the instructor begins to discuss an idea or procedure that you are unable to follow.

Your instructor has just lectured the material, but have you read the material in the textbook yourself yet? Reading can make a world of difference in how difficult the material seems to you. When reading a textbook, remember that it is not a novel. Each paragraph – sometimes even each line – contains deep ideas. It may take you 20 minutes or longer just to absorb and understand a single page. That is normal. Read it with blank paper available and a pencil in your hand. Pause, look away, and look again at the book. Work through the examples yourself, until you thoroughly understand each step. *Writing things down is far more effective than highlighting or underlining.*

Students who collaborate can develop a synergy among themselves that supports their learning, helping them to learn more, more quickly, and more lastingly.

Talk to your instructor and to other students. With determination and a positive outlook – and a little help – you will accomplish things you once thought impossible.

2.3   Rules for problem solving

In graduate school I wrote lessons for myself how to solve problems. These are my notes.

1. Ask yourself, *"What's the problem* (or difficulty)?" Check that it is correctly stated. Be sure you are not working on the wrong problem.

2. Start from the *beginning*. This sounds easy, but often students start from the middle of an idea, and so get confused.

3. It you do not know how to start, *pause, relax*, and study it. Jot down possible ideas. Put your *full attention* on each detail of the theory and problem.

4. *Check* everything. This does not mean to do it again. If you did it again, you did it again, but did not check it. Check each detail separately. For example, check numerical addition and signs separately.

5. *Smile*. If you are happy and cheerful, you will think better, do better work, and get better grades.

2.3.1   Example of asking what is the problem

Here is a problem from the text:

Today is your first day driving a city bus. When you leave downtown, you have twenty-three passengers. At the first stop, three people exit and five people get on the bus. At the second stop, eleven people exit and eight people get on the bus. At the third stop, five people exit and ten people get on. How old is the bus driver?

The rule is to ask what is the problem. When I call on students to read and try to solve the problem, they start reading, "Today is your first day driving..." I stop them mid-sentence. Wrong! One must first state the problem. This means skipping ahead, and to start reading, "How old is the bus driver?" Once you know the problem, you can then start reading from the beginning, "Today is your first day driving...". Now that you know the problem, the age of the driver, and you know that you are the driver, you know the answer. The answer is your age. There is no need to waste time reading the rest of the problem.

You may have to translate the problem from English to mathematics. This is an important step, which many unfortunately overlook. Knowledge requires understanding concepts, words, and the math symbols.

Rule 2 is to start from the beginning, which is only after you know what the problem is.

Speaking about starting from the beginning, remember Julie Andrews in *The Sound of Music*, singing "Let's start from the very beginning... a very good way to start..." ♪♫

2.3.2   Relax, and full attention

How to relax? Look away for a moment, tell yourself, *"Think of nothing."* Look again at the material. Repeat, looking away, thinking of nothing. Practice this doing homework. Do this during tests in order not to be nervous.

When working, you must give your full attention to the task, and do one step at a time. If you do several steps, you are not giving your full attention to each step. Enjoy listening to music, but not when working. Do your studying without music, then stop and enjoy the music.

Smiling helps one relax, and therefore if you smile you will understand the material better.

2.3.3 How testing improves memory

I read an interesting article that I wish to share.

http://www.sciencenews.org/view/generic/id/64316/title/
How_testing_improves_memory

November 6th, 2010; 178 #10 (p. 16)

Covering up the words and quizzing yourself is a better learning strategy than repeatedly reading the words, psychologists reported in the Oct. 15
Science. Self-testing strengthens the memory by creating keywords as clues for retrieving the word pairs later on.

Students are often lulled into a false sense of knowledge by staring at information. "The illusion is, you read something and think you'll remember it. But if you don't try to retrieve it, you don't know if you know it.

Students often neglect it while preparing for exams. Students will underline or highlight facts they think are important, but before test time they only reread the marked information. They think they know it because they have read it so many times, but they haven't practiced the skill they'll need on the test, which is retrieval.

2.4 Inductive and deductive reasoning

Deductive reasoning is mathematics. We start with some assumptions, apply rules, and arrive at conclusions. Inductive reasoning is not mathematics, for we assume something based upon several observations, and the conclusion may or may not be true. This violates the consistency requirement. Why then do we discuss inductive reasoning, including topics like successive differences, in a mathematics course? The answer is that since mathematics is a human creation, people do not suddenly create a complete consistent mathematical system. Instead, the creative process advances in stages, starting with guesses and hypotheses. Sometimes a conclusion based upon inductive reasoning can be developed into deriving the result by deductive reasoning.

In addition, inductive thinking, while not being completely rational, does reduce the chances of making a wrong decision. Of course, it does not eliminate the chances of a wrong decision.

2.5   Frog climbing up a well

Here is a problem that confuses students, but illustrates an important principle of mathematics.

A frog is at the bottom of a 20-foot well. Each day it crawls up 4 feet, but each night it slips back 3 feet. After how many days will the frog reach the top of the well?

Students focus on the fact the frog slipped each night, while ignoring the condition at the end. When the frog reached the top of the well, it did not slip back. After 16 days, it was 4 feet short of the top. The next day it crawled up 4 feet and got out. The answer is 17 days.

The principle is to look at the boundary. Boundary conditions are very important in mathematics and physics, but will not be part of the course. As an example, listen to the sounds of a violin. All the strings are identical, except that they have different lengths. At the ends, the strings do not vibrate, and this makes each string sound differently.

We mention things which are not part of the course, and for which you will not be responsible. The reason is to give you a feel of the boundaries. I consider myself as a tourist guide, with you students hiking on the trails. I may mention a beautiful mountain on the side of a trail. You may look at and enjoy the mountain, but are not advanced hikers to actually hike up the mountain. The idea is that no human being can know all there is to know about mathematics, and so we can only learn some things, not everything. However, it would not hurt to mention some advanced topics so that you can appreciate some limits on you studies.

2.6   Critical thinking

Students need to learn not only the basic principles of mathematics and science, but also how to process and analyze the information they receive. Without the ability to think critically and independently, students, and thereby the general public, become susceptible to misinformation and fall prey to the fallacies of junk science.

Throughout our elementary and secondary school training, we have been conditioned to accept most of the information we have been taught as factual. Current practice tends to emphasize results rather than processes, to rely heavily on memorization of facts rather than the understanding of concepts.

2.7 The induction postulate

This is a postulate that we can use to prove mathematical relations. Suppose we want to prove something like $\sum_{i=1}^{n} g(i) = f(n)$. That is, we want to prove that for a sum of some terms we get a certain result. Mathematical induction says we can prove this if we can show two things as true. One thing is $g(1) = f(1)$.

The second thing we need to show is this. If we assume

$$\sum_{i=1}^{n} g(i) = f(n) \text{ implies } \sum_{i=1}^{n+1} g(i) = f(n+1),$$

then our original statement is true by the induction postulate.

Here is a simple example. We want to show that the sum of the first odd numbers with the last number n is n^2. In this case, $g(i) = (2i - 1)$ and $f(n) = n^2$. That is $\sum_{i=1}^{n} (2i - 1) = n^2$.

The first thing we show is that $g(1) = f(1)$, that is, $1 = 1$.

We then assume $\sum_{i=1}^{n} g(i) = n^2$ and prove that $\sum_{i=1}^{n+1} g(i) = (n+1)^2$.

The proof is very easy.

$$\sum_{i=1}^{n+1} g(i) = \sum_{i=1}^{n} g(i) + g(n+1).$$

Substituting $g(i) = (2i - 1)$:

$$\sum_{i=1}^{n+1} (2i - 1) = \sum_{i=1}^{n} (2i - 1) + (2(n+1) - 1)$$

$$= n^2 + 2n + 1$$

$$= (n+1)^2.$$

This means that the sum is true for all n.

Suppose we made a mistake, and assumed $\sum_{i=1}^{n}(2i-1) = n^2 + 1$.

We want to see if $\sum_{i=1}^{n+1}(2i-1) = (n+1)^2 + 1$.

Expanding:

$$\sum_{i=1}^{n+1}(2i-1) = (n^2+1) + \big((2(n+1)-1)\big).$$

Expanding the right hand side we get

$$n^2 + 1 + 2n + 1$$

$$= (n+1)^2 + 1.$$

This seems to satisfy the induction postulate. Or does it? What did we forget? Did we prove a false statement? If so, the postulate is not valid!

We forgot to examine the case for n=1. We have $1 = 2$ which is false. The postulate is fine!

2.7.1   Sum of first n odd natural numbers

Here is a proof of a theorem to get the sum of the first natural numbers where the last number is n. The sum is n^2. The proof is easy, but instructive as it illustrates the use of symbols.

The theorem is:

$$\sum_{i=1}^{n}(2i-1) = n^2.$$

The capital sigma means sum, where i starts out with 1 and ends with n.

Another way of writing this is

$$\sum_{i=1}^{n}(2i-1) = 1+3+5+\cdots+(2n-1)$$

We need to familiarize ourselves with the sigma notation.

To prove this, we will us a mathematical principle called *induction*. This says that if a statement is true for $n = 1$, and if we prove assuming it is true for n implies it is true for $n + 1$, it is then true for all n.

Imagine climbing up a ladder. You know that you can get started. You also know that if you are at any rung you can get to the next rung. Therefore, you know for all values of the ladder, going up to infinity, it will work.

Here is the proof based upon induction.

It is clear that it is true for $n = 1$.

We assume it is true for *n:* $\sum_{i=1}^{n}(2i-1) = n^2$.

We will prove it is true for $n+1$:

$$\sum_{i=1}^{n}(2i-1) + (2n+1) = (n+1)^2.$$

If this is true, it is true for all *n.*

Simplify:

$$n^2 + 2n + 1 = n^2 + 2n + 1.$$

The first term, n^2, is just the induction assumption.

This is true, and so it is true for all *n* by the induction assumption.

2.7.2   Sum of first n numbers - Gauss

Let us ask a simpler question. What is the sum of the first n numbers? That is, find $\sum_{i=1}^{n} i$.

We can add the first and last numbers to get $n+1$. Let us add the second number with the second to the last number. The second number is 2, and the second to the last number is $n-1$. We added 1 to the first element of the pair, and subtracted 1 from the second element. This shows that the value of the second pair is the same as the value of the first pair. The sum is also $n+1$. We can pair the numbers up. For n even, we have $\frac{n}{2}$ pairs, with each pair having the same value $n+1$. The sum is $\frac{n}{2} \cdot (n+1)$. For example, suppose $n = 4$. We have two pairs, the value of each is 5. The sum is 10.

For n odd it is a bit tricky, but we get the same result. We also have $\frac{n}{2}$ pairs, with the value of each $n+1$. The middle pair is just a half a pair. For $n = 3$, we have two pairs. The first pair is equal to 4. The middle number is 2, a half a pair. We have $\frac{3}{2}$ pairs. Multiply by the value of each pair, 4, gives the sum 6.

The answer, for both n even and n odd is

$$\sum_{i=1}^{n} i = \frac{n}{2} \cdot (n+1).$$

This was first developed by Carl Friedrich Gauss.

2.7.3 Sum of first n numbers by induction

Another proof of this is by induction. This is instructive as it is an easy example of using induction.

To prove the sum of the first n numbers is

$$\frac{n}{2}(n+1), \text{ that is } \sum_{i=1}^{n} i = \frac{n}{2}(n+1).$$

We will use induction. We assume the theorem is true, and look at the $n+1$ case.

First, if $n=1$ the theorem is obviously true.

$$\sum_{i=1}^{n+1} i = \sum_{i=1}^{n} i + (n+1)$$

$$= \frac{n}{2}(n+1) + (n+1)$$

We have a common factor $n+1$:

$$= (n+1)\left(\frac{n}{2}+1\right)$$

$$= (n+1)(n+2)\frac{1}{2}.$$

Assuming it is true for n, we find it is true for $n+1$.

This proves it by induction.

2.7.4 Sum of the cubes of the first n natural numbers

We can use the same logic to find the sum of the cubes of the first n natural numbers:

$$\sum_{k=1}^{n} k^3.$$

We can prove this by induction. The result is

$$\sum_{k=1}^{n} k^3 = \left(\sum_{k=1}^{n} k \right)^2.$$

We want to prove

$$\sum_{k=1}^{n} k^3 = \left(\sum_{k=1}^{n} k\right)^2 = \left(\frac{n(n+1)}{2}\right)^2$$

$$\sum_{k=1}^{n+1} k^3 = \sum_{k=1}^{n} k^3 + (n+1)^3.$$

We need to prove that

$$\sum_{k=1}^{n+1} k^3 = \left(\frac{(n+1)(n+2)}{2}\right)^2$$

or

$$\left(\frac{n(n+1)}{2}\right)^2 + (n+1)^3 = \left(\frac{(n+1)(n+2)}{2}\right)^2$$

$$(n+1)^3 = \left(\frac{(n+1)(n+2)}{2}\right)^2 - \left(\frac{n(n+1)}{2}\right)^2$$

$$= \frac{(n+1)^2}{4} \times \left((n+2)^2 - n^2\right)$$

$$= (n+1)^2 \left(n^2 + 4n + 4 - n^2\right)/4$$

$$= (n+1)^3.$$

2.8   Variables

Another point we need to understand is the concept of a *variable*.

The important thing is that the units on both sides of an equation must agree, for otherwise we made a mistake. This is called *dimensional analysis*. This is a simple and very powerful tool.

Here is an example of a problem using equations and variables:

The gas gauge on a vehicle read 1/8 full. The driver added 15 gallons, and now the gauge read 3/4 full. How much more gas does the driver have to add to fill the tank?

The first step is to read the problem. When we read a problem, we do not start reading "The gas gauge on…", but start reading the problem "How much more gas…" After reading the problem, and defining a variable or two, we read the problem again from the start to get given data. Students tend to read problems from the start of the problem, which is not the best way.

The problem is to find how much gas. We define a variable. Let x gal be the number of gallons the driver has to add. Now we read from the beginning. "The gas gauge on…" We define another variable, f gal, the number of gallons in a full tank. Reading the first two sentences of the problem, we have

$$\frac{1}{8} f \text{ gal} + 15 \text{ gal} = \frac{3}{4} f \text{ gal}$$

This equation has 3 terms. A *term* is an expression near a + sign. *Factors* are things multiplied together in a single term. Each term has a factor gal. The first term,

$$\frac{1}{8} f \text{ gal}$$

has three factors: $1/8$, f, and gal. The factor *gal* is a common factor to all three terms, and so we can cancel it from each term:

$$\frac{1}{8} f + 15 = \frac{3}{4} f.$$

We simplify. Note that we simplify before proceeding:

$$\frac{3}{4} f - \frac{1}{8} f = 15$$

$$\frac{5}{8} f = 15$$

$$f = 24.$$

Now we read the next sentence:

"How much gas…" We get the equation, after canceling the common factors gal:

$$\frac{3}{4} f + x = f.$$

We simplify and get

$$x = \frac{1}{4} f.$$

Using the first equation, $f = 24$, we get $x = 8$.

Here is how students usually write it:

x is gal added, f is full tank.

This is not mathematically correct, for x is a variable, not a physical quantity, gallons of gas. We must say $x \; gal$ is the number of gallons the driver has to add. There are three concepts here.
- One is the idea of number, which is 15.
- The other is units, which is gallons here.
- The third is the idea of variable, which are x and f.

We must keep these three things separate. In Europe, gas is measured in liters. Suppose the problem was the driver added 5 gallons from a jerry can. How many more liters are needed? Errors arise when these things are not kept separate.

The concept of variable is very difficult to explain to students, because they are so used to saying things like x is gal added instead of x gal is the number of gallons to add. Students need to practice doing problems to overcome this incorrect notion.

Another approach is to define one variable, and to do the rest mentally. This is not easier, and is error prone. However, high school books tend to favor this approach, incorrectly thinking it is easier.

2.8.1   Student error

A student wrote:

Let g be an abbreviation for gallon. We have

$$\frac{1}{8}f + 15g = \frac{3}{4}f.$$

Wrong! f is a variable, and does not include the units g. The equation should be

$$\frac{1}{8}fg + 15g = \frac{3}{4}fg.$$

Now we can simplify by cancelling g from each term.

$$\frac{1}{8}f + 15 = \frac{3}{4}f.$$

With the wrong equation, we cannot cancel anything.

2.8.2   Word problems

This is discussed in the book in the chapter of basic concepts of algebra.

To do word problems, we first read to find what the problem is. We then read from the beginning. We use the given to define variables. We will get equations. The number of equations must match the number of variables. E.g., if we have 3 variables, we need 3 equations to be able to solve the problem.

Here is an example of a problem:

A project calls for 3 pieces of wood. The longest piece must be twice the length of the middle-sized piece, and the shortest piece must be 10 inches shorter than the middle-sized piece. If the three pieces are to be cut from a board 70 inches long, how long can each piece be?

Let L inches be the length of the longest piece. Note that we have to write the units, inches. Let M inches be the length of the middle-sized piece, and S inches be the length of the smallest piece.

The first sentence translated into mathematics is:
L inches $=2M$ inches.

The next is:
S inches $= M$ inches $-$ 10 inches

The last sentence is:
L inches $+ M$ inches $+ S$ inches $= 70$ inches

We now have 3 equations for the 3 unknown variables.

The first step is to simplify. We can cancel the units, inches, from each term. Our equations become:
$L = 2M$
$S = M - 10$
$L + M + S = 70$

We solve by the usual methods of algebra which you should know. This is discussed below in the section "Solutions of linear equations" in the chapter "Problem solving".

An important principle is doing one step, that is, one thing, at a time. One of the things that help is to visualize the problem. This means you stop working, visualize, and then continue working. Do not visualize at the same time as working (one step at a time). E.g, this problem. Visualize the pieces. Then you would not have erred saying the largest is 21 and the middle 26, as some did.

Note the importance of writing the units, inches. E.g, for motion problem we may be given the speed as 50 mph. This is 50 miles per hour. The word "per" means divide. The speed is 50 miles / hour. If the car went for 2 hours, how far would it go? We do not need to memorize formulas, but simply write the equation with the proper units:

$$\frac{50 \text{ miles}}{\text{hour}} \times 2 \text{ hours.}$$

The unit "hour" cancels, and we get the answer 100 miles.

Had we made a mistake and divided by 2 hours instead of multiplying, the hours would not have cancelled.

2.8.2.1   One variable

Students say they prefer to use one variable rather than several. For example, let us look at the problem of finding lengths of pieces of wood. The textbook by Miller on page 295 states the following:

Let x represent the length in inches of the middle-size piece.
$x =$ *the length of the middle-size piece*

Since this has an equal sign, it is an equation. Since the left-hand side is a variable, the right hand side is also a variable. In the business world, variables are often phrases rather than single letters as students are accustomed.

The next line in the text is

$2x =$ *the length of the longest piece*

The left-hand side is a variable, and so is the right. The next line

$x - 10 =$ *the length of the shortest piece*

Note that there are four variables:
x
the length of the middle-size piece
the length of the longest piece
the length of the shortest piece

We are given that the total length is 70 inches. The equation using variables is:

the length of the longest piece
$+$ *the length of the middle-size piece*
$+$ *the length of the shortest piece* $= 70$

Substituting the definitions of the variables in terms of x:
$2x + x + (x - 10) = 70$

Note that although it looks like we are working with one variable, we are actually working with several variables. IMHO it is easier to write the variables in the order given in the problem, defining each explicitly.

3. Set theory

Mathematics is not just numbers, but ideas and concepts. We will start with a new concept that most of you are unfamiliar with. This is the concept of a set. A set is a bunch of different numbers, colors, letters, people, anything, except possibly other sets. To say it more precisely, ***a set is an unordered collection of distinct elements.***

Let's start talking about sets of numbers because you are used to thinking of mathematics as numbers.

Here are some examples.
$A = \{1, 2, 3, 4, 5\}$, $B = \{1, 2, 3\}$.

Here are more examples:
C={red, white, blue}, D={current Rider students}

In mathematics, we can say what we wish as long as there are no contradictions. Russell showed that there are contradictions with set theory, such as the Barber Paradox. However, if we do not permit sets of sets, we do not get the contradictions. In other words, a set is a collection of any type of objects, except other sets. Actually, to avoid a paradox, we cannot permit a set to contain itself as an element.

Mathematicians like to generalize. Can you think of a generalization for sets of numbers? We can have sets of fractions. We could have a set of all natural numbers {1,2,3,...} with no end, as we know there is no largest number.

3.1   Counterexamples - things that are not sets

A set is an unordered collection. To clarify, we gave some counterexamples, of mathematical entities that are not sets.

(a,b) is an *ordered pair*, also called a *vector*. This is written using parentheses.
A vector has one row and k columns. If k is 2, it is an ordered pair. If k is 3, it is an ordered triplet.

We can generalize this, as mathematicians like to generalize. Instead of one row, we can discuss n rows.

An object with n rows and k columns is an n by k *matrix*. This is written using brackets. We specify each element of a matrix by its row and column. It is like specifying an address giving a street and avenue. With sets, we do not specify the address, as order does not matter.

We will discuss ordered pairs briefly, and not discuss how to add or multiply vectors or matrices. We will not discuss matrices in this course.

We mention these things to show what are the boundaries of the course, that is, what we will not discuss.

3.1.1 Pictures of a set and a matrix

It helps to see pictures. Here is a picture of a set. There are 8 balls in a bag. We can shake the bag, making the balls move, but it will be the same set. Below the picture of a set is a picture of a matrix. The 8 balls are in an ice cube tray. If we rearrange the balls, we have a different matrix.

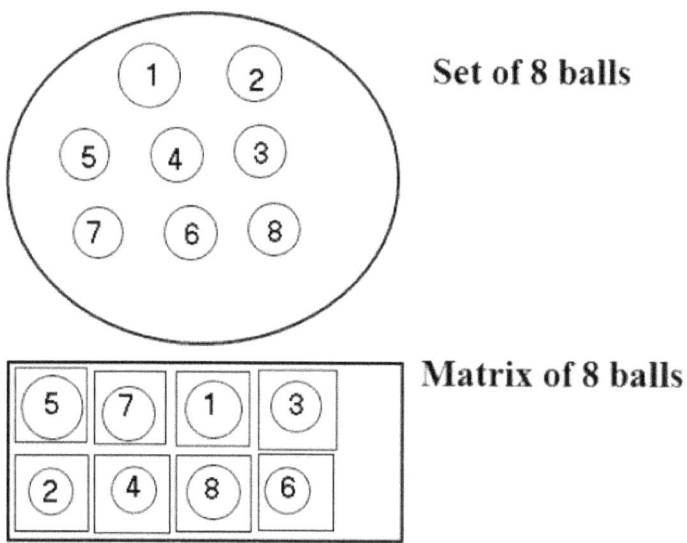

Set of 8 balls

Matrix of 8 balls

3.2 Operations on sets

Before we continue with more examples, let us talk about operations.

How can we add $A = \{1, 2\}$ and $B = \{1, 2, 3\}$? One way is to have the new set to have every element in the old sets, either in one or the other. This would be $C = \{1, 2, 3\}$. Another way of adding sets would be the new set has elements that are in both, elements in one and in the other. This would be $D = \{1, 2\}$.

Well, we have two ways of adding. One is *union*, which has elements that are in the first *or* in the second. This is the set C. We write this as $A \cup B = C$. The other is *intersection*, which has elements in the first *and* in the second. This is the set D. We write this as $A \cap B = D$.

Note that union uses the word OR, and intersection uses the word AND.

Let A be the set of stores on 42nd St. in Manhattan, and B be the set of stores on Fifth Ave. The union would be the set of stores on 42nd St. or 5th Ave. The intersection would be the set of stores on 42nd St. and 5th Ave.

The ideas of union and intersection are easy. The only confusing part is how not to get these two symbols mixed up. Here is my mnemonic trick. You can remember the symbol for union thinking of the preacher stretching his arms out like a U, saying, "You are all welcome!" The intersection reminds one of the stern boss who puts his arms down and yells, "You all better do this right!"

3.2.1   Set subtraction, the empty set

Set difference, or subtraction. Now that we know how to add sets (union or intersection), let us see how we would define subtraction. Let us look at numbers and try to generalize.

$5 - 3 = 2$. We put 5 sticks down, remove 3, and are left with 2.
Let us define $A = \{1,2\}$ and $B = \{1,2,3\}$.
Let $B - A = E$. The set E consists of the elements of the set B, with the elements of the set A removed.

$$E = \{\cancel{1}, \cancel{2}, 3\} = \{3\}.$$

The rule is difference of two sets is a set with the elements of the first set with the elements of the second set removed.

Let us look at the number 0. We know that 2 - 2 = 0. Put down 2 sticks, remove the sticks, and we have nothing left. What has happened is something interesting. Numbers like 2 or 3 are called *natural (or counting) numbers.* These numbers represent physical objects. We can have 2 sticks or 2 balls or 2 people. The concept of zero is a new mathematical concept. Recall that mathematics is a human creation, and so mathematicians created the idea of zero, letting 0 be a number. Numbers like 0,1,2, and such are called *whole numbers.* We can create new mathematical concepts as long as it does not lead to a contradiction. Again, 0 is a whole number but not a natural number.

We have the same thing with sets. $\{1, 2\} - \{1, 2\} = \{\}$.

Let E be the set of elephants in a zoo, and let B be the set of baby animals in the zoo. $E - B$ is the set of elephants excluding the baby elephants. The fact that B includes baby tigers does not affect E. Sometimes it helps understanding if you think of people or animals, instead of just letters and numbers.

We want to say that the subtraction of two sets is a set. We therefore create the idea that the *empty set*, a set of no elements, is a set. This sounds strange when students hear it for the first time. A student asked me the following: "If a set is a collection of objects, how can we say the empty set is a set, when it is not a collection of objects?" The answer is that mathematicians generalize the concept of set to include the empty set. Let us recall that the number 0 was also strange when people heard it for the first time.

There is another way to understand the idea of the empty set. We assume that given any two sets, we can subtract one from the other and get a set. This means that if the two sets are the same, set subtraction gives the empty set, which must be a set. The existence of the idea that the empty set is a set follows from the assumption that subtraction any two sets gives a set.

Thanks to the student for asking me that question! When students ask question, we all benefit with increased understanding!

Note that mathematics does not have to make sense, but merely has to be consistent. An empty set does not make sense, but is required for consistency. We will give more examples of where mathematics does not make sense.

We have a special symbol for the empty set: $\{\} = \emptyset$. Note that the empty set is not $\{\emptyset\}$. This is technically the set containing the empty set. However, we cannot define a set of sets as inconsistencies arise.

3.2.1.1 Student errors

If A and B are sets, $A - B$ are elements in A and not in B. Some students said $A - B$ are elements in B and not in A. We need to pay attention to these type of errors.

$A \sim B$ means these sets have same cardinalities. It does not mean that the sets are similar, as this is set theory, not geometry.

3.2.2 Multiplication of sets, ordered pairs

Multiplication of sets.   When we multiply numbers, we do repeated addition. $5 \times 2 = 5 + 5 = 10$. We define set multiplication differently. Since this was developed by the French mathematician Descartes, we call this the Cartesian product. The Cartesian product of sets A and B is the set of all possible ordered pairs of elements, the first from A and the second from B.

Note that in order to define the product we had to introduce a new mathematical concept, *ordered pairs*. A pair of shoes is an ordered pair, as there is a left and a right shoe. A point on a graph is an ordered pair, the x coordinate and the y coordinate. We write this using parentheses: (x, y). Contrast this with writing sets using braces {}.
Be sure you do not confuse these symbols!

Here is what happened in class one day:

"I do not know what an ordered pair is!" said a student.

"You know what a pair is," I replied.

"Yes."

"Good. Now what does ordered mean?"

"Okay, we write a, b in order. How do we write it so that it means an ordered pair?

We write it as *(a,b)*."

"Now, class, now that you know what an ordered pair is, give me an example of a set of ordered pairs."

"I don't know!"

$A \times B$

We have to learn how to think logically, step by step. When we see two words, look at the *last* one: pair. There are lots of pairs, just as there are a lot of Aranoff's. Look at *ordered* pair. This makes one type of pair, just as Sanford Aranoff is one of the Aranoff's. We do not start by looking at the first word: *ordered*, but the last one: *pair*.

You know $A \times B$ is a set of ordered pairs. You should be able to think conversely - give an example of a set of ordered pairs. When you study you reviewed $A \times B$, and it is a set of ordered pairs. In addition, you should look at the same thing from another viewpoint: Find a set of ordered pairs.

Look at a concept, and think of an example. Then pick an example, and ask yourself which concept it is. E.g., think about this one. Let L be the set of your left shoes at home, and R the set of right shoes. Let S be the set of your pairs of shoes.

Is $S = L \times R$? Explain.

Answer: No. S is the set of pairs. $L \times R$ is the set of all pairs of left and right shoes. If you had two pairs of shoes, a brown pair and a black pair, $L \times R$ would include a left brown and a right black as one of the pairs, while this unmatched pair is not in the set S. Note that $n(S) = 2$ while $n(L \times R) = 4$.

Cute thing we said in class. Let M be the set of men in the class, and W the set of women. The set $M \times W$ is the set of all possible couples (m, w). This shows that the Cartesian product consists of very different things than the set. The set consists of students, and the product set consists of couples.

3.2.2.1　Student errors

$\{1,2\} \times \{1\} = \{(1,1),(2,1)\}$.

There are two elements. Some made the error of adding (1,2) to have 3 elements.

Another error was writing $(1,2)$ as $\{1,2\}$, writing a set instead of an ordered pair.

3.2.3   More on Cartesian products

A point in a plane can be written as an ordered pair (x, y). The horizontal distance of the point from the origin is x, and the vertical distance is y. Since the points are ordered pairs, the set of points in a plane is $R \times R$. Likewise, points in space can be written as an ordered triplet (x, y, z). The set of points in space is $R \times R \times R$. A point is an ordered pair plus an origin, another ordered pair. A point is a vector.

Distance. The distance between two points in a plane is defined using the Pythagorean theorem. If x is the horizontal distance and y the vertical distance, we define $d^2 = x^2 + y^2$. Likewise, the distance between two points in space is $d^2 = x^2 + y^2 + z^2$.

This does not have to refer to space. Suppose we wanted to collect information on weight (w), height (h), and shoe size (s) from each of a random sample of individuals. The data would consist of ordered triplets (w, h, s).

Mathematicians like to generalize. We can define the set of points in 4 dimensions as $R \times R \times R \times R$. It does not matter that space is really 3 dimensional and not 4 dimensional. Mathematics does not have to correspond to reality. However, suppose we wanted to look at show size, and include age (a). The data would consist of ordered elements like (w, h, s, a). The results of mathematics can apply to many different things.

Why stop at 4 dimensions? Why stop? Actually, as long as we have consistency, we can continue. We can continue forever: $R \times R \times R \times \cdots$ to get an infinite dimensional space. However, we do get into trouble unless we take a subset of this space, restricting ourselves only to points that have a finite distance. This is called a Hilbert space. This is not part of the course.

You may wonder who cares about an infinite dimensional space, wondering if this has anything to do with your life. The answer is that this is critical for the development of quantum mechanics, the major theory of physics in the 20th century. Quantum mechanics is responsible for computers, the Internet, drug development, just to mention a few. You see this idea that mathematicians did just for fun, to see how far they can push the logic, lead to world-shaking events.

3.2.3.1   Products of other mathematical entities

A number times a number is always a number.

A vector times a vector can be defined as a vector or a number. There are two definitions of a vector product.

A matrix times a matrix is a matrix, but only certain matrices can be multiplied.

Again, contrast this with sets. A set times a set is a set. A set of what? A set of ordered pairs of elements from each set.

3.2.4   Unary operations on sets

Unary operations. The above discussion dealt with operations on two sets giving a third set. There are two unary operations. One changes a set into a whole number. This is called the *cardinality* of the set. The only thing hard about the word cardinality is the word itself. The cardinality of set A is written as $n(A)$. For example, if $A = \{1,2,3\}$, then $n(A) = 3$.

Do not confuse a set $\{\}$ with the cardinality of a set $n(\{\})$.

Another unary operation is the complement, A'. This is the set of all elements in the *universe* which are not in A. The universe is the set of all elements under consideration. Mathematically, it is just a set.
For example, if
$U = \{1,2,3,4,5,6,7,8,9,10\}$, and $A = \{1,2,3\}$,
then $A' = \{4,5,6,7,8,9,10\}$.
The first unary operation, $n(A)$, changes a set into a number.
The second unary operation, A prime, written as A', changes a set into another set.

Note that the prime operation is that the prime character appears after the letter. Contrast this with the cardinality operation, $n($, that appears before the letter.

Question: What operation takes a set into a number? Answer: cardinality, $n(\ )$.

To summarize: There are two unary operations on sets. One takes a set into a number. This is the cardinality of the set. The other operation takes a set into a set. This is the complement.

3.2.5 Equality and equivalence of sets

Equality. $A = B$ if every element in A is in B, *and* every element in B is in A. For example, $\{1,2,3\} = \{1,3,2\}$. Note that there are two conditions, connected by *and.* Here is an example where one condition is met, but not the other. Let $A = \{1,2\}$ and $B = \{1,2,3\}$. Every element in A is in B, but not every element in B is in A. In this case, $A \neq B$, and we say A is a subset of B, writing it as $A \subseteq B$. (Note that a set is a subset of itself).

If the cardinalities of two sets are equal, but the sets are not equal, we say the sets are *equivalent.* For example, let $A = \{1,2,3\}$, and $B = \{1,2,4\}$. In this case, $A \neq B$, but $A \sim B$. The tilde means equivalence. It means $n(A) = n(B)$.

The notation for subset is similar to the notation for inequality. E.g., $3 \leq 5$. We note that we can correctly write $5 \leq 5$. If we want inequality but wish to clarify that they are not equal, we write it as $3 < 5$, without the line under the inequal sign. For sets, if we want to say that A is a subset of B, but that A is not equal to B, we write it as $A \subset B$, again without the line under the subset sign. E.g., $\{1,2\} \subset \{1,2,3\}$ means $\{1,2\}$ is a *proper subset* of $\{1,2,3\}$.

Note that $\{1,2,2,3\} = \{1,2,3\}$, as all the elements in the left set are in the right set, and vice-versa. This means that if we wrote $\{1,2,2,3\}$, we would have to cancel the extra 2, in a fashion similar to simplifying fractions. This stresses the idea that sets are collections of distinct elements. That is

$$\{1,2,2,3\} = \{1,2,\not{2},3\} = \{1,2,3\}.$$

3.2.6   Formal definitions of set operations

For you to properly understand something, you must know both the informal and formal definitions. The formal definition is the definition using the proper mathematical symbols. The informal definition is an English sentence or two that makes sense to you.

Here are some formal definitions.

Definition of set equality:

$A = B$ if $\forall a \in A \rightarrow a \in B$ AND $\forall a \in B \rightarrow a \in A$.

The $\forall$ means pick any a that is an element of the set A, it will be an element of the set B.

Note that we need two conditions.

If $A = \{1, 2\}$ and $B = \{1, 2, 3\}$,

the first condition is satisfied but not the second.

Definition of subset:

$A \subseteq B$ if $\forall a \in A \rightarrow a \in B$.

The $\forall$ means that you pick a until you are done, and if $a \in A$, etc.

Suppose $A = \varnothing$. Pick a until we are done. There is nothing to pick.

Therefore, the condition is satisfied, and $\varnothing \subseteq A$ for all sets A.

E.g., $\{1, 2\} \subseteq \{1, 2, 3\}$.

Definition of proper subset:

$A \subset B$ if $A \subseteq B$ AND $A \neq B$.

Note that there are two things that must be true.

Set equality theorem.

If $A = B$ then $A \subseteq B$ and $B \subseteq A$.

Proof:

If $A \subseteq B$ then $\forall a \in A \rightarrow a \in B$.

If $B \subseteq A$ then $\forall a \in B \rightarrow a \in A$.

This is the definition of equality.

The definition refers to elements.

The theorem refers to subsets.

Equivalent sets.

We say A is equivalent to B if $n(A) = n(B)$.

We write this as $A \sim B$.

E.g., $\{1, 2\} \sim \{1, 3\}$.

3.3   Cardinal Number Formula

The number of elements in the union of two sets, that is, the number of elements in either one set or the other set, is equal to the number of elements in each set minus the number of elements in both sets. We subtract, as we count this twice.

The cardinal number formula is, where A and B are sets:

$$n(A \cup B) = n(A) + n(B) - n(A \cap B).$$

It is called the cardinal number formula because a cardinal number is a number that represents cardinality of sets. A cardinal number can be a whole number or a transfinite number (used for cardinalities of infinite sets).

Let A be the set of students in the first row, and B be the set of students in the second row. Let one student move her chair between the first and second rows, so that she can be counted as being in either row. Counting students in either row, we have to subtract this one student, as we counted her twice.

The importance of this classroom exercise is that when students see things related to their own bodies they comprehend it better.

3.3.1   Examples using Venn diagrams

The numbers in the Venn diagram below represent cardinalities. For example, the cardinality of the set A is 24 + 4, or 28. The numbers represent the regions indicated, not the cardinality of the set.

Find the cardinality $n(A \cap B')$.

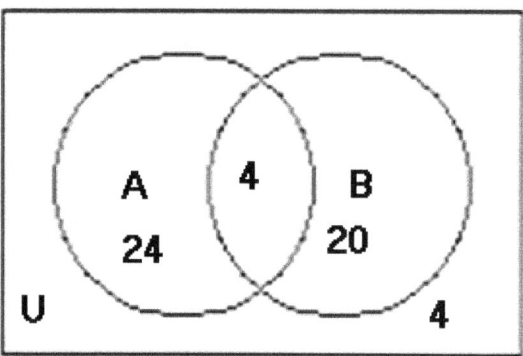

The answer is that it must be in A AND it must not be in B.
The answer is 24.

This confuses students.
A mistake would be to give the answer as 4.
The error is $n(A \cap B) = 4$, confusing B with B'.

Another error is finding the union instead of the intersect.
This confuses $\cup$ with $\cap$.

We must pay attention to *possible* errors in all the things we do. While working, we know that we are likely to confuse the union and intersect signs. Since the complement sign is so small, we may confuse a set with its complement. Likewise, we often confuse plus signs with minus signs. Therefore, we must pay extra attention to these types of errors while working on a problem.

3.3.1.1 Another example

Shade the regions representing the set $A' \cap B'$.

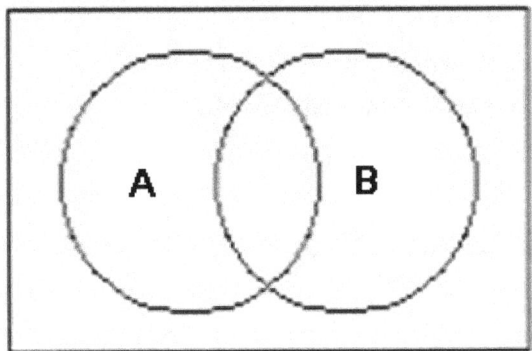

Here is the answer. My shading is crude, but you get the idea.

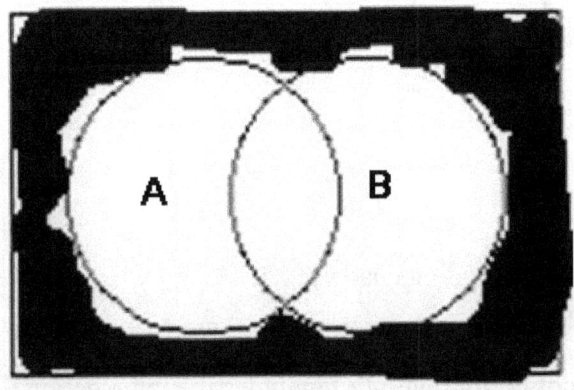

Every shaded region must satisfy two criteria:
It must not be in A, *and* it must not be in B. The intersection symbol means AND.

3.3.1.2 A third example

Draw an appropriate Venn diagram for the sets A and B and use the given information to fill in the number of elements in each region.

$n(A) = 19$, $n(B) = 13$, $n(A \cup B) = 25$, $n(A') = 11$

Answer: We see that $n(A) + n(B) - n(A \cup B) = 7$.

Using the cardinal number formula, we get $n(A \cap B) = 7$.

This is the number of elements in the intersection.

This is the Venn diagram.

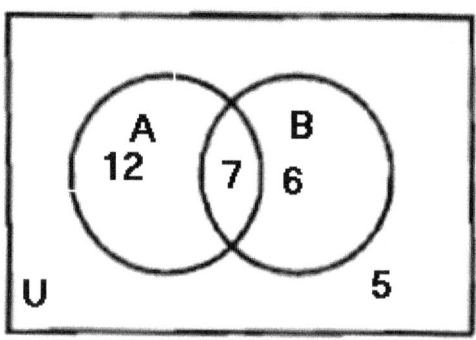

Note that $12 + 7 = 19$ and $6 + 7 = 13$.

Instead of 12, 7, 6, some students had 19, 25, 13, confusing $\cup$ with $\cap$.

Another error was not noting that n(A) includes the intersection.

3.4   De Morgan's Law

Here is the proof. The proof is to show that two sets are equal.
Instead of proving equality by saying all the elements in one set are in
the other, we prove equality by saying the first is a subset of the
second and the second is a subset of the first.

Proof of De Morgan's Law:
Theorem: For any two sets S and T, $(S \cap T)' = S' \cup T'$.

Let us define sets A and B as:
$A = (S \cap T)'$, $B = S' \cup T'$.

To prove $A = B$, we will prove $A \subseteq B$ AND $B \subseteq A$.

First we will prove $A \subseteq B$.

Let $x \in A$.

$\therefore x \notin S$ or $x \notin T$

$\therefore x \in B$

Since $\forall x \in A, x \in B$

$\therefore A \subseteq B$

To prove the second statement $B \subseteq A$:

Let $x \in B$

$\therefore x \in S'$ or $x \in T'$.

$\therefore x \notin S$ or $x \notin T$.

$\therefore x \notin (S \cap T)$

$\therefore x \in (S \cap T)'$

$\therefore B \subseteq A$

3.5 Subsets

Subsets of $\{1,2,3\}$ are $\{\},\{1\},\{2\},\{3\},\{1,2\},\{1,3\},\{2,3\},\{1,2,3\}$.
Note how we wrote them down systematically. First with no elements, then with one element, etc.

Note furthermore that as we add elements to a set, the number of subsets double. For example, if we look at the subsets of $\{1,2,3,4\}$, we have all the subsets of $\{1,2,3\}$, and then all these subsets with the element 4 added.

If a set has n elements, the number of subsets is 2^n.

If $n = 0$ the number of subsets is $2^0 = 1$.

3.6 Surveys

One example of using set theory is surveys. Here is a typical problem:

A survey of 240 families showed that
        91 had a dog;
        70 had a cat;
        31 had a dog and a cat;
        91 had neither a cat nor a dog nor a parakeet;
         7 had a cat, a dog, and a parakeet.
How many had a parakeet only?

We solve this by first drawing the Venn diagram.

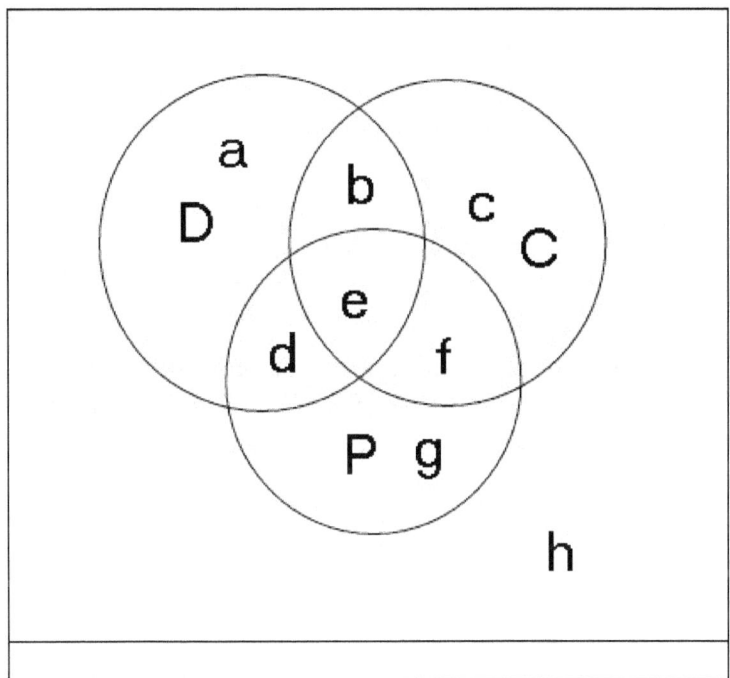

We use capital letters for sets. Let D be the set of people who own dogs, C be the set of people who own cats, and P the set of people who own parakeets. These sets are circles. The universal set is the rectangle.

We use lower case letters for numbers in various regions. We label regions $a, b, \ldots, h$.

Region a is the number of people who own dogs only. Region b is the number of people who own dogs and cats. These regions are actually sets. Set a is the set of people who own dogs only, whereas set D is the set of people who own dogs and maybe other pets.

We use the given information to write equations.
91 had a dog gives $a + b + d + e = 91$.
70 had a cat gives $b + c + e + f = 70$.
31 had a dog and a cat gives $b + e = 31$.
91 had neither a cat nor a dog nor a parakeet gives $h = 91$.
7 had a cat, a dog, and a parakeet gives $e = 7$.
240 people in the survey gives $a + b + c + d + e + f + g + h = 240$.

We have 6 equations and 8 unknowns. The problem was not to find all the unknowns, but only the number who had a parakeet only. In general, we need 8 equations if we wish to find 8 unknowns.

We substitute and get
$a + d = 60$

$c + f = 70 - 31 = 39$

We note that 240-91=149:

$c + f + g = 149 - 91 = 58$

This is the number of animals − number of dogs.

$g = 58 - 39$

$g = 19$

The answer is g, the number who had a parakeet only, which is 19.

This problem unfortunately involves many equations and so is time consuming to solve. There are any number of ways of solving these equations. I just happened to do this.

3.6.1 Another way to solve surveys

The above solved the survey by writing down simultaneous algebraic equations for unknown variables and solving the equations.

We can solve this another way, but looking at sets and using the various relations we know about sets.

Let $B = D \cup C$ and $F = D \cup C \cup P$

$n(P) = n(F) - n(B)$ from the Venn diagram.

$n(F) = n(U) - n(F')$

$n(U) = 240$

$n(F') = 91$

$\therefore n(F) = 149$

$n(P) = n(F) - n(B).$

$n(B) = n(D) + n(C) - n(D \cap C)$

$= 91 + 70 - 31$

$= 130$

$\therefore n(P) = 149 - 130$

$= 19.$

To summarize, we had to find two intermediate results from the given.

We also had to use the formula for the cardinality of the union of two sets.

3.7 Infinite sets of numbers

Let us discuss infinity.

It was fun when I gave candies to two of my grandchildren, a three-year old and a five-year old. I made piles in two plates. I asked them which pile had more. The three-year old started counting, but could not get past 15. The five-year old said she could count more. I said how can we find which has more without counting. The answer was by taking one candy from each plate until one plate is empty. When this happened, the five-year old said, "I won!" She had less candies.

We can count piles of candies. We cannot count numbers, because we never finish.

Which is more, the set of natural numbers $N = \{1, 2, 3, \ldots\}$ or the set of whole numbers $W = \{0, 1, 2, 3, \ldots\}$? We play the candy game. Match the 1 from the natural numbers to the 0 from the whole numbers, then match the 2 from the natural numbers to the 1 from the whole numbers, and keep on matching. We never finish. No one can say "I won!" Therefore we cannot say one set has more elements. That is, $n(N) = n(W)$. This is surprising, as the set W contains one element, 0, that the set N does not have. We need to think about this very surprising aspect of infinity.

One property natural numbers have is that given any number, regardless how large, we can add 1 and get a larger number. There is no largest number. There is no end. This is the meaning of infinity, not finite, no end.

In mathematics we can make any statements we want to providing we do not get into a contradiction. Cantor made the statement that the cardinality of the set of natural numbers is a number. He made up a number, which he is allowed to do. He called this aleph null, and wrote $n(N) = n(W) = \aleph_0$.

Aleph is the first letter of the Hebrew alphabet, and the first letter of the Ten Commandments in Hebrew. It is a silent letter. One of the names for God in Hebrew is "Ein-Sof", which literally means no end, or infinite. Philosophers have had problems with infinity, saying that since God is infinite, nothing else can be infinite.

One of my sons-in-law said that there is a limit to the number of numbers, as there is a limit to the number of particles in the universe. I replied that indeed there is a finite number of particles. If we match each number to a particle, we will get a finite set which is a subset of the actual infinite number of numbers. Mathematics, as opposed to science, does not have to correspond to reality, nor does it have to make sense. We only require consistency.

3.7.1   Aleph Null

Let us not confuse $\aleph_0$ with ∞.

Aleph-null is a number, the cardinality of the set of natural numbers. We use lazy eight in writing things like
$1, 2, 3, \cdots, \infty$.

This means the numbers keep on going on and never stopping. Lazy eight merely indicates it does not end, but it is not a number. Aleph-null is a number. If you say the cardinality of the set of numbers is ∞, you are wrong, as ∞ is not a number.

Let us stress that we do not say, as we sometimes tend to say, that infinity means it goes on forever. We are giving a precise definition, the number of elements in the set of natural numbers.

Now that we know that $\aleph_0$ is a number,
let us look at some of its properties.
One is that $\aleph_0$ is larger than any natural number.
Another property is $\aleph_0 + 1 = \aleph_0$.
We give a name for this type of number: *transfinite.*

We will see below that there are other transfinite numbers.
For example, the cardinality of real numbers (decimals) between 0 and 1 is larger than $\aleph_0$, and is equal to a transfinite number we call c.

What is interesting about transfinite numbers is that these are larger than any natural number, and yet not infinity.

Let us try to generalize this, as mathematicians do. Let us look at very small numbers larger than zero. For example, $0.0001 < 0.001$. We can get smaller and smaller numbers by adding zeros to the right of the decimal. We can define infinitesimal numbers as numbers smaller than any number but larger than zero. Infinitesimals are important for calculus, and beyond the level of this course.

3.7.2   Sets of fractions

We can count 1, 2, 3, 4, but there is no end to the natural numbers. How about fractions, that is, rational numbers? Which is more? We play the candy game. We use Cantor's proof, and find that we can match each natural number with a rational number, and go on forever. Since we never finish, we cannot say which is more.

Let N be the set of natural numbers, and Q the set of rational numbers.
We can say $n(N) = n(Q) = \aleph_0$.

Another way of saying this is that the sets N and Q are countable, meaning that the cardinalities are aleph null.

The discussion of the cardinality of the set of rational numbers could be clarified by saying that since a rational number is an ordered pair of two numbers, there is a one-to-one correspondence between rational numbers and integers. Note the abstraction of rational numbers to ordered pairs. The fraction 3/4 is essentially the ordered pair (3,4). The beauty of abstraction is that normally when we think of fractions we do not think of ordered pairs!

The decimal expansion of a rational number is eventually periodic (in the case of a finite expansion the zeroes which implicitly follow it form the periodic part). This is another answer to the question about 1/3 = 0.33333... An irrational number is not periodic.

Given any two rational numbers, we can find another rational number greater than the smaller and less than the larger. We merely add the two numbers and divide by 2. This fact is called *density*. It is often referred to as the "density property". However, it is not strictly a property, for the word property means a postulate, an arbitrary assumption. We speak about the closure property, for example, where this is an assumption that does not lead to contradictions.

Consequently, rather counterintuitively, while the rational numbers look like they are a continuous set, they are actually countable. This is very strange! This is another example how the logic of mathematics leads to things that do not make sense!

3.7.3   Names of number sets

These are names of sets of numbers:

N Natural   Positive numbers, not including 0.
W Whole   Includes 0
Z Integers   Includes negative numbers
Q Rational   Integers and fractions
R Real   Decimals
C Complex
P Prime numbers

Real numbers are called real numbers, as they are not imaginary numbers.

In addition, there are *transfinite* numbers and *infinitesimal* numbers.
We will not be discussing sets of these numbers.

Some students error saying natural numbers include zero. There is nothing natural about zero. There is a long history about the concept of the number zero. Many people were puzzled, asking "How can nothing be something?" The same confusion is true for the empty set, discussed above. Natural numbers, on the other hand, can be identified with objects that we see.

3.7.4 Cardinality of sets of real numbers

The cardinality of the set of real numbers between 0 and 1 is larger than the cardinality of the set of natural numbers. The text gives Cantor's proof. Here is another way of understanding this.

Look at all the points on a line from 0 to 1. There are an infinite number of points. Proof: Take any 2 points, e.g., 0.234 and 0.235. I can always find a third point half way between, by adding and dividing by 2: 0.2345. Question: Which is larger - the number of natural numbers or the number of points between 0 and 1?

Answer: We play the candy game. We take point 2 and match it with $\frac{1}{2}$, point 3 match it with $\frac{1}{3}$, point 4 match it with $\frac{1}{4}$, and go on forever. We use all the numbers and yet there are points between $\frac{1}{2}$ and $\frac{1}{3}$, between $\frac{1}{3}$ and $\frac{1}{4}$. Therefore, the number of points on a line is larger. One infinity is larger than the other is.

We give a name to this infinity: c. This stands for continuous. We think of the line as a continuous set of points. $\aleph_0 < c$.

Of course, we can get the same cardinality by looking at the number of points on a line between 60 and 61. It surprises me that students get confused with this on exams.

How many subsets are there to the set of natural numbers? The rule is the number of subsets is equal to 2^n, where n is the cardinality of the set. The number of subsets of the set of natural numbers is

$2^{\aleph_0}$, as $\aleph_0$ is the cardinality of

the set of natural numbers.

Mathematicians have proven that this is equal to the cardinality of the number of real numbers between 0 and 1. That is,

$$2^{\aleph_0} = c.$$

We will not discuss this complicated aspect of set theory, as it is beyond the level of the course.

3.7.5   Examples of infinities from physics

Students may be thinking that who cares about these two infinities, sets of numbers that contain so many elements that it takes forever to count them. Here is an example from physics where the distinction between the countable infinity aleph-null and the continuous infinity c is important.

The old theory of the atom was that an atom is a miniature solar system, with electrons orbiting the nucleus. If we hit an atom and knock an electron to a higher orbit, it gives light as it falls down to its original orbit. We imagine that we can get any color we want, depending on how hard we hit the atom. We make an experiment using a simple atom such as helium. We expect to find all the colors of the rainbow. Surprisingly, we find only certain colors. Electrons can only be in certain orbits, not anywhere in-between. Quantum mechanics explains this. We can make a one-to-one correspondence between the energy levels (the orbit radii) and the natural numbers. In the old theory, the correspondence between energy levels matches the real numbers.

Go to http://hyperphysics.phy-astr.gsu.edu/hbase/quantum/atspect.html to see beautiful pictures of helium spectra. We get lines of color, not continuous colors like the rainbow. This proves that the energy levels correspond to natural numbers, not real numbers. For example, students in a school building can be on the first floor, the second floor, or the third floor, but not on a floor between the first and second.

4. Number theory

Ask a student what is mathematics, and invariably she will respond saying mathematics is numbers. Well, having discussed the abstract idea of sets, let's discuss numbers, which is what many think mathematics is.

This chapter deals with natural numbers:
$1, 2, 3, \cdots$.

4.1 Divisors and divisibility

A divisor of a natural number is a natural number that divides the number with no remainder. Actually, since this chapter deals with natural numbers only, we can rewrite it:

A divisor of a number is a number that divides the number with no remainder.

That was the informal definition. Here is the formal definition:
a is **divisible** by b if $a, b \in N$ and $\exists k \ni k \in N$ and $a = bk$.
In English:
a is divisible by b if there is a number k such that $a = bk$.
In this case, b is a divisor of a.
Also note that b and k are factors of a.

For example, divisors of 4 are 1, 2, and 4.

The next question is divisibility. Here we deal with digits. Digits are parts of numbers. E.g., the number 12 has two digits. Think of the digits in an automobile odometer.

There are various rules. For example, a number whose last digit is 0 is divisible by 10. The proof is very easy. Multiply any number by 10 and the product has a 0 on the end.

Likewise, a number whose last digit is 0 or 5 is divisible by 5. To prove this, multiply any number by 5, and we get a number whose last digit is 0 or 5.

Divisibility by 2 is easy also. Since we know that 10 is divisible by 2, any number that ends with a 0 is divisible by 2. If the last digit is divisible by 2, the number is divisible by 2. This is because if we divide by 10, which is divisible by 2, and the remainder is divisible by 2, the number is divisible by 2. E.g., 386. Divide by 10 and get 38 with a remainder 6:
$386 = 38 \times 10 + 6$.

Likewise, since 100 is divisible by 4, any number that has the last 2 digits divisible by 4 is divisible by 4. In addition, since 1000 is divisible by 8, any number that has the last 3 digits divisible by 8 is divisible by 8.

We note that $9 + 1 = 10$. Therefore, $9 \times 2 = 20 - 2$, and
$9 \times 3 = 30 - 3$, etc. This shows that if the sum of the digits adds to 9 the number is divisible by 9.

Using similar thinking, we can show that if the sum of the digits adds to 3 the number is divisible by 3.

4.1.1   Divisibility by 7

There is a divisibility test for 7. It is complicated, and so not very useful.

Let n be the number. Write n as $10x + y$, where $y < 10$.

If n is divisible by 7, so it $20x + 2y$.

We can write this as

$20x + x - x + 2y$

$= 21x - x + 2y$.

Since 21 is divisible by 7, so is $2y - x$.

Subtract twice the last digit from the rest of the number (without the last digit).

See if this is divisible by 7. We may have to apply the test again.

Example: 91
$9 - (1 \times 2) = 7$.

Another example: 8281
$828 - 2 = 826$
$82 - 12 = 70$

4.2   Prime Numbers

A prime number is a number that is not divisible by anything but 1 and itself. The importance of prime numbers is because of the **Fundamental Theorem of Arithmetic.** This states that any composite number (a number that is not prime) can be written as a product of primes in a unique fashion.

For example, $12 = 3 \times 4 = 2 \times 6 = 2^2 \times 3$.

Using composite numbers there are two ways to write the factors. Using prime numbers, 2 and 3, there is only one way.

What this means is that we have another way of picturing numbers.

Instead of picturing 12 as 12 sticks laid on the ground, or 12 pieces of chalk, we can picture 12 as the product of its prime factors. It is a new way of looking at something very familiar. Mathematicians love to find new ways of looking at objects.

Here is an example of the extension of math to our social lives. Prime factorization is unique. In addition, each person is unique.

One way to find a list of prime numbers is Eratosthenes' sieve. See www.faust.fr.bw.schule.de/mhb/eratclass.htm. This is cute.

Since every number can be factored in a unique fashion using prime numbers, the divisibility tests greatly help in finding these factors. We try small numbers like 2, 3, 10, and such, and then repeat with the remainder. The divisibility tests speed up this process. It helps to remember the divisibility tests.

4.2.1 Cardinality of the set of prime numbers

How many prime numbers are there? Let P be the set of prime numbers. What can we say about $n(P)$?

Is there a largest prime number? The book proves that if we assume there is a largest prime number, we get a contradiction. You can find any prime, and I can find a larger prime. We can use matching to find $n(P)$. We match natural numbers to prime numbers: 1-2,2-3,3-5,4-7, etc. We never finish matching. Therefore, the cardinalities of the sets are the same.

The proof is like this. Suppose there were a largest prime number p. Form a number m which is a product of all the prime numbers less than or equal to p, and add 1. For example, suppose the largest prime number is 7.

In this case $m = 2 \times 3 \times 5 \times 7 + 1 = 211$.

In this case 211 is not divisible by any number ≤ 7, and is larger than 7.

Let N be the set of natural numbers: $N = \{1, 2, 3, \cdots\}$.

Let I be the set of integers: $I = \{\cdots, -1, 0, 1, \cdots\}$.

Let P be the set of prime numbers.

$P \subset N$. Every prime is a number, but not every number is a prime.

$n(P) = n(N) = n(I) = \aleph_0$.

There is another way to look at the infinite number of primes. We know that every number can be written as a product of prime numbers. If there were a finite number of prime numbers, once we got to large numbers, we would have no more prime numbers left for the unique factorization.

Suppose we asked which of your classes this semester had the most students. Simple. You count the students in each class, and see which is the largest. Mathematically, we look at the sets of students in each class, and see which has the largest cardinality.

For prime numbers, we have a problem. We cannot count the number of prime numbers, for there is no end. There is always another larger prime number. We use matching. We match natural numbers to prime numbers: 1-2,2-3,3-5,4-7, etc. We never finish matching. Therefore, the cardinalities of the sets are the same. With real numbers between 0 and 1, we can match natural numbers to real numbers: 1-1/2, 2-1/3,3-1/4, etc., and use all the natural numbers and still have real numbers left. Therefore, there are more real numbers than natural numbers.

4.2.2   More comments about prime numbers

The following topics are made for your general interest, as the detailed explanations and proofs are beyond the level of the course.

Encryption is based on prime numbers – two prime numbers to be exact. When multiplied together, two prime numbers will yield a product that is only divisible by one and itself - and those two prime numbers. These prime numbers are used in a complex algorithm to scramble (encrypt) a message or file. Thereafter, the two prime numbers are needed again in order to unscramble (decrypt) the message or file.

Size of the Prime Numbers – The size of prime numbers used dictate how secure the encryption will be. A message encrypted with prime numbers that are 5 digits in length (40-bit encryption) yields about 1.1 trillion possible results. However using 128-bit encryption (16 digit numbers) yields this number of possible results:
340,282,366,920,938,463,463,374,607,431,768,211,456

Time Needed To Crack - Mathematically speaking, based upon today's top computing power 40-bit, and 128-bit encryption could be broken in 1 second and 11,000 quadrillion years, respectively. This is why 128-bit encryption is the standard used world wide to protect financial transactions and sensitive data.

The largest prime number as of 2011 has 13 million digits.

4.3   GCF, LCM

One application of prime numbers is finding common factors to pairs of numbers. Given two numbers, we would like to find the common factors. If there is a common factor, we would like to find the greatest common factor. This is **GCF**. Define F first - factor. Then C - common to two numbers.

There are three ways to find the GCF: prime factors method, dividing by prime factors, and subtracting (or dividing) one from the other (Euclidean algorithm).

We can generalize from factors to multiples. A multiple of a number is the number times a number. E.g., 100 is a multiple of 10. 10 is a multiple of 10. Given two numbers, we want to find common multiples. We want the smallest of the common multiples, for the multiples increase forever. This is the **LCM**.

E.g., it takes me 30 sec to swim a lap. A student swims it in 20 sec. We start swimming together. After how long will we start together again? We find the LCM, which is 60. After 60 sec, I will have swum 2 laps and she will have swum 3, so that we start together again.

Some students made errors with finding the GCF of 135 and 216. One error is getting 1080. This number is a multiple, not a factor. The confusion was with LCM. Some got 3 for the GCF.. Yes, 3 is a CF, a common factor. However it is not the GCF.

4.3.1   Euclidean algorithm for GCF

One way of finding the greatest common factor of two numbers is the Euclidean algorithm. The way it is usually presented is dividing the larger by the smaller, and continuing to divide the smaller by the remainder. It can also be done by subtracting the smaller from the larger, and usually (but not always) subtraction is easier.

My proof of the Euclidean algorithm.
We wish to find the GCF of two natural numbers a and b.
A factor common to the two numbers will also be a factor to the difference.

Look at the GCF of $(a - b), b$.

That is, subtract the numbers to get another pair.

Repeat, subtracting the smaller from the larger

until we get a number $<=$ the smaller of the original numbers.

This number is the largest common factor.

We can write a computer program that you can program
into your programmable calculator.

function $GCF(a,b)$

// Given two natural numbers it finds the Greatest Common Factor

while $a > b$ do

   if $a := a - b$ then

      $a := a - b$ // This means replace a with $a - b$.

Like the STO button on a calculator.

   else if $a = b$ then

      break

   else begin

      $d := b$; // Store b temporarily

      $b := b - a$; // Make b smaller.

      $a := d$; // a is the original b before we changed it.

   end;

$GCF := a$

E.g, find GCF of 81 and 216.

$216 - 81 = 135$ We subtract the smaller from the larger.

$135 - 81 = 54$ We take the smaller two numbers,

                        and subtract the smaller from the larger.

$81 - 54 = 27$

$54 - 27 = 27$

GCF is 27.

Another example: 455, 130
$455 - 130 = 325$
$325 - 130 = 195$
$195 - 130 = 65$
$130 - 65 = 65$
GCF is 65.
To check:
$455 \div 65 = 7$
$130 \div 65 = 2$

Another example: 168, 90
$168 - 90 = 78$
$90 - 78 = 12$
$78 - 12 = 66$
$66 - 12 = 54$
$54 - 12 = 42$
$42 - 12 = 30$
$30 - 12 = 18$
$18 - 12 = 6$
$12 - 6 = 6$
GCF is 6.
In this case, the division approach is faster.
We do not have to divide 216 by 81, but we can subtract, which is easier.

4.3.2   GCF × LCM = mn

Given two natural numbers m and n, we can prove the following theorem:

$GCF \times LCM = mn$.

Proof.

Let us write $m = p^a q^b r^c$ and $n = p^d q^e r^f s^g$,

where all the variables are natural numbers, and p, q, r, and s are prime numbers. This is perfectly general, for all numbers can be written as products of prime numbers.

Let $a \leq d$ and $b \leq e$ and $c \geq f$.

$GCF = p^a q^b r^f$, as we take the smaller exponents.

$LCM = p^d q^e r^c s^g$, as we take the larger exponents.

$GCF \times LCM = p^{a+d} q^{b+e} r^{f+c} s^g = mn$.

This proves the theorem.

Another way of writing the theorem is $LCM = \dfrac{mn}{GCF}$.

It is an easy of getting the LCM of a product of two numbers if we know the GCF.

4.4 Golden ratio

This is based upon drawing a rectangle, and then a smaller rectangle inside the larger rectangle, where the two rectangles are *similar* (same shape, different size).

A golden rectangle is a rectangle whose ratio of length to width has a certain value, so that we can draw a similar rectangle inside, and keep on drawing. Similar: ratio of sides is the same. When you were a child, your shape is similar to your present shape, as it is the same shape, just a different size.

See http://goldennumber.net/face.htm

This is related to Fibonacci numbers. See http://www.sciencenews.org/view/feature/id/8732/title/ Mathematical Lives of Plants

The Fibonacci numbers tend to crop up wherever the golden ratio appears, because the ratio between two consecutive Fibonacci numbers happens to be close to the golden ratio. The larger the two Fibonacci numbers, the closer their ratio to the golden ratio.

4.5 The first decade

Consider a line with integers marked. The markings are 0,1,2, ..., and -1, -2, ... If we look at point 2 and go back one, we get to point 1. Go back one, we get to point 0. Go back one, and we get to point -1. The first decade of non-negative integers is 0,1,2,3,4,5,6,7,8,9. The next integer, 10, requires another digit. Likewise, if we look at the non-positive integers, we see the first decade is 0,-1,-2,-3,-4,-5,-6,-7,-8,-9.

When historians matched years to integers, they did not include a zero. This inconsistency leads to confusion. Look at the year 2 AD. Go back one year, and we get 1 AD. Go back another year, and we get 1 BC (or 1 BCE as some historians write it). This is an inconsistent mathematical relationship. This leads to confusion as to the end of the first decade. If we say there must be 10 years in a decade, the last year of the first decade is 10 AD. This violates the rule that a new decade is when another digit is needed. Likewise, we have confusion as to the start of the second century AD, whether it is 100 AD or 101 AD.

Historians have tended to consider 101 AD as the start of the new century. However, with current computer usage this inconsistency is unacceptable. EDI, Electronic Data Interchange, requires a new digit to mean a new decade. EDI is the transfer of structured data, by agreed message standards, from one computer system to another without human intervention. In this formal standard, the year 2000 AD is the first year of the 21st century. The fact that we need to change all the digits going from 1999 to 2000 means 2000 is the start of the new century. We need not be troubled by the fact that the first century AD has 99 years, for it actually has 100 years if we include the zero required for consistency.

5.   Real numbers

Now that we discussed natural numbers, we wish to discuss all forms of finite numbers, namely, real numbers. We will briefly discuss complex numbers. We will not discuss transfinite and infinitesimal numbers.

5.1   Rational and Irrational Numbers

A rational number is a fraction. If we write a rational number as a decimal, some terminate. For example, ½ written as a decimal is 0.5. Others repeat. For example, 1/3 written as a decimal is 0.333.... What happens when we divide 1.0 by 3, we get 3 with a remainder 1. Bring down a 0, divide by 3, we get again 3 with a remainder 3. This never ends.

A terminating decimal can be viewed as a repeating decimal, with the digit 0 repeating.

If we try to write 1/7 as a decimal, when we divide by 7 we get a remainder, the value of which is always < 7. This means there are only so many digits in the quotient before it starts repeating.

If a decimal never repeats, then it cannot be written as a fraction, as we have just shown that all fractions can be written as repeating decimals. A decimal that never repeats is an irrational number.

Some wrong answers by students:

"A rational number is a fraction, and an irrational number is a negative number."

Well, if so, then $-\dfrac{1}{2}$ is rational and irrational.

"Irrational number - when the denominator is 0." The denominator is *never* 0.

These errors are caused by not pausing and checking after writing the answers.

5.2   Properties of numbers

Properties of Addition and Multiplication of Real Numbers

Let R be the set of real numbers.

Let $a, b,$ and $c \in R$. $a, b,$ and c are real numbers.

	Addition	**Multiplication**	**Sets** (examples)
Closure:	$a + b \in R$	$a \times b \in R$	$A \cup B$ is a set.

Words: Add (multiply) two numbers, get a number.

Commutative: $a + b = b + a$ $a \times b = b \times a$ $A \cup B = B \cup A$

$$A \times B \neq B \times A$$

Associative : $(a+b) + c = a + (b+c)$ $(ab) \times c = a \times (bc).$

Identity : $\exists 0 \in R \ni a + 0 = a$ $\exists 1 \in R \ni a \times 1 = a$ $A \cup \varnothing = A$

Words: There exists a number such that any number
added (multiplied) to this number gives the number.

Inverse : $\forall a \, \exists (-a) \ni a + (-a) = 0$ $\forall a \neq 0 \, \exists \dfrac{1}{a} \ni a \times \dfrac{1}{a} = 1$

Words: For any number there is another number
called the inverse such that
the number and the inverse is the identity number.

Distributive : $a \times (b+c) = a \times b + a \times c$

Note that this is for both addition and multiplication.

Example. Use the inverse property to prove $-(-a) = a$.

Let us take the inverse of $-a$.

This inverse is a number by the inverse property.

Using the inverse property on $-a$, we have

$$-a + (-(-a)) = 0, \text{ or}$$

$$-(-a) = a, \text{ proving it.}$$

Here is a counterexample of closure. My little granddaughter cannot count past 15. If we ask her what is 8+3, she knows it is 11. If we ask her what is 8+8, she does not know. Her number system does not satisfy closure.

The identity and inverse properties are assumptions that a number exists. For example, the identity property for addition is not satisfied for the set of natural numbers.

5.2.1   Why we cannot divide by 0

Division is the inverse of multiplication. Given a number a, there is an inverse b such that $a \times b = 1$.
If a is 0, then there is no number that we can multiply by it to get 1.
That is why we have to restrict division by numbers not zero.

Since this is very important, whenever we have a denominator, we must explicitly write that the denominator is not zero.

Another example of the need to explicitly state that the denominator is not zero is with lines and slopes. The equation of a line is
$Ax + By = C$. If $B \neq 0$, we can divide every term by B.
This gives us the usual slope-intercept equation for a line.

However, if $B = 0$, we cannot divide by B. The line is $Ax = C$.

There is no slope. The concept of slope does not exist. As a line becomes more vertical, the slope becomes larger. However, it is wrong to say the slope of a vertical line is infinite. There is simply no slope.

5.3　McKay's theorem

The book mentions this. I tried to prove it. Here is how I did it.

Given natural numbers a, b, c, d such that $\dfrac{a}{b} < \dfrac{c}{d}$.

We can find a rational number between these two rational numbers.

Theorem :

$$\frac{a}{b} < \frac{a+c}{b+d} < \frac{c}{d}.$$

Proof :

$cb > ad$

$\therefore cb + ab > ad + ab$

$b(a+c) > a(b+d)$

$\dfrac{a+c}{b+d} > \dfrac{a}{b}$

Likewise :

$cb > ad$

$\therefore cb + cd > ad + cd$

$c(b+d) > d(a+c)$

$\dfrac{a+c}{b+d} < \dfrac{c}{d}.$

This sometimes is easier than adding the numbers and dividing by 2.

5.4   Percentages

Big words in mathematics and science are often made up of smaller words with no spaces. The word "percent" is made of two parts: per and cent. Per means divide, and cent means 100. Percent means divide by 100.

4% of $100 is $4.

Percent change. If something changed from a to b, the percent change is

$$\frac{b-a}{a} \times 100.$$

Note that we always divide by the original value. Students sometimes err and divide by the final value. If $b > a$, the percent change is positive.

5.5   Rounding

We round to the nearest value, trying to minimize the error.

E.g., let us round to tenths.

$$3.43 \approx 3.4$$
$$3.46 \approx 3.5$$

Let us try to understand the logic. When we round, we make an error. We wish to minimize the error. If we round 3.43 to 3.4, the error is 0.03. If we round to 3.5, the error is 0.07.

How do we round 3.45? We get the same error if we round to 3.4 or to 3.5. Teachers often say to round up to 3.5. However, in mathematics, we try not to do things without a reason. Actually, rounding up is not always fair. Suppose I am the bank. If you owe me money, we will round up; if I owe you, we will round down. This is not fair! We have to be fair, and rounding up is not fair!

We round to even:

$$3.45 \approx 3.4$$
$$3.55 \approx 3.6$$

In one case we rounded down, the other we rounded up. This is fair!

Why not round to odd? $3.55 \approx 3.5$? If so, we will never get a 0, and this will cause errors if we round again.

E.g., $3.005 \approx 3.01$ or 3.00. We take the latter.

Note that 3.00 is not the same as 3. The extra zeros tell us the accuracy.

5.6 Order of operations

Multiplication is like super addition, and exponentiation is like super multiplication. The order is first do parentheses, then the most super first. Don't worry about Aunt Sally. This means parentheses, exponentiation, multiplication and division (the order does not matter), and finally addition and subtraction (the order also does not matter).

5.7   Complex numbers

Complex numbers

Let us assume that the equation $x^2 = -1$ has a solution.

There is no real number that satisfies this equation.

Let us invent a number, and call it i, where $i^2 = -1$.

Since these assumptions are consistent with the rest of mathematics, we can accept them.

Let a be a real number. $a \in R$, where R is the set of real numbers.

We can define numbers ia. These are called **imaginary numbers**.

Let $b \in R$. We can then write numbers like $a + ib$.

Numbers like $a + ib$ are called **complex numbers**.

The set C of complex numbers can be written as $C = R \times iR$.

A complex number is an ordered pair (a, ib).

There is a theorem which states that every equation in one variable, no matter how complicated, can be solved using complex numbers. 4We therefore do not need a new type of number which would be an ordered triplet, such as (a, ib, jc).

Powers of imaginary numbers :

$i^3 = -1 \times i = -i$

$i^4 = (i^2)^2 = (-1)^2 = 1$

Since we need two quantities to define a complex number,
we can identify complex numbers as points on a plane,
with the x-axis the real numbers and the y-axis imaginary numbers.

There are more things that we can say.
I'll just write them, but not hold you responsible.

$(a + ib)(a - ib) = a^2 + b^2$. Just expand it, and you'll see.

The distance from a point in the complex plane to the origin
is called the **absolute value**.

$|a + ib| = \sqrt{a^2 + b^2}$.

Here is something that we will not prove. It is Euler's formula:

$\cos\theta + i\sin\theta = e^{i\theta}$

This equation is one of the basic equations needed for engineering,
such as power generation and transmission.

Can you evaluate Euler's formula for $\theta = 0$?

For $\theta = \pi$, we have $e^{i\pi} = -1$.

This incorporates the special mathematical quantities e, i, π, and -1.
This is cool!

5.8   Exponents

Here are rules for exponents. They are very easy.

$$2^2 \times 2^3 = (2 \times 2) \times (2 \times 2 \times 2) = 2^5.$$

This is obvious for natural numbers.

We postulate (make an arbitrary assumption) that this is true for all numbers, including real and complex. Therefore

$$2^{2.5} \times 2^{3.7} = 2^{(2.5+3.7)} = 2^{6.2}.$$

Likewise, since $\left(2^{\frac{1}{2}}\right)^2 = 2^1 = 2$, we can say that $\sqrt{2} = 2^{\frac{1}{2}}$.

What about 2^0 ? We apply the addition rule:

$$2^0 \times 2^3 = 2^3.$$

Cancel the 2^3. We get $2^0 = 1$.

Likewise negative exponents:

$$2^{-3} \times 2^3 = 2^0 = 1.$$

Therefore $2^{-3} = \dfrac{1}{2^3}.$

What about the exponent of an exponent?

$$(2^2)^3 = (2^2) \times (2^2) \times (2^2) = 2^6.$$

We see that we multiply the exponents.

If we multiply two different numbers raised to the same power, we can take the product to that power:

$$a^m b^m = (a \cdot b)^m.$$

The rules for real numbers as exponents are postulates, arbitrary assumptions upon which mathematics is based.

5.8.1   Logarithms

Subtraction is the inverse of addition. By subtracting, we undo what we just did.

E.g., $5 + 3 = 8$, and $8 - 3 = 5$.

Likewise, division is the inverse of multiplication.

A logarithm is the inverse of exponentiation. E.g.,

$$10^2 = 100 \text{ and } \log_{10} 100 = 2.$$

Raising 10 to the power 2 gives 100. In order to get the 2 back from the 100, take the log 100.

So many students are fearful of logarithms! Remember, mathematics is not magic! If we do something, we need to have an operation to undo what we did.

Once I took my children to the Museum of Natural History. There were signs there saying the earth is billions of years old. My daughter said she does not know what a billion is. I said you know what 9 is. A billion has 9 zeros. Log 1 billion = 9.

6.   Functions, systems of equations

We start with simple ideas, and move to more complicated ideas. We start with the simple, and generalize. This is the fun part of mathematics.

6.1   Points

Point. In one dimension, we need one number, such as x. In 2 dimensions, a point is an ordered pair: *(x,y),* such as in Manhattan: (St, Ave). Give the pair, we know the location of the point.

2 points. New thing: *Distance* between the points.
We cannot discuss distance when talking about one point.
E.g, one point a house on 1 Jardin St., and the other a house on 2 Jardin St.

One dimension: Distance between two points x_1 and x_2 :

$$d = |x_2 - x_1|, \text{ or } d^2 = (x_2 - x_1)^2$$

E.g., $x_1 = -4$, $x_2 = 3$. We get $d = 7$.

2 dimensions: **Pythagorean theorem**.
Let d_1 be the horizontal distance between two points,
and d_2 be the vertical distance. The distance d is
$$d^2 = d_1^2 + d_2^2.$$

Note that it is easier to write the distance squared than to write the distance d using the square root sign.

HW. Draw a right triangle, on side horizontal, the other vertical. Measure the three sides.
Compare with the Pythagorean theorem. Note your errors. E.g., if

one side is 3 (cm or in), the other 4, and you measure the hypotenuse to be 4.95 the error is 0.05, and the relative error is 0.05/5.

If we write the radical (square root) sign, we have

$$d^2 = (x_2 - x_1)^2 + (y_2 - y_1)^2$$

$$d = \sqrt{d^2} = \sqrt{(x_2 - x_1)^2 + (y_2 - y_1)^2}$$

Note that $\sqrt{a^2 + b^2} \neq \sqrt{a^2} + \sqrt{b^2}$

E.g.: $a = 3, \quad b = 4$

$$\sqrt{3^2 + 4^2} = 5 \neq \sqrt{3^2} + \sqrt{4^2} = 7$$

6.1.1   Midpoint

Given two points we can connect them with a straight line. Therefore, two points define a line. We will discuss lines further below.

Given two points, we can define a third point, the midpoint.

In one dimension, it is the average location, which is the sum of the locations divided by 2.

In two dimensions, the midpoint (x,y) is such that x is the average location of the x coordinates and y is the average location of the y coordinates. That is, the midpoint of

$$(x_1, y_1), (x_2, y_2)$$

is

$$\left(\frac{x_1 + y_1}{2}, \frac{x_2 + y_2}{2} \right).$$

What is important is that the midpoint is a *point*, not a *number*. Again, given two points, we get a third point, the midpoint.

6.2   Circles

If we have 2 points (h,k) and (x,y),

let the distance between them be r.

The distance formula gives

$$r^2 = (x-h)^2 + (y-k)^2$$

A circle is a set $\{(x,y) \mid (x-h)^2 + (y-k)^2 = r^2\}$.

This is a set of points. *A circle is a set.*
A set of what?
Answer: A set of points.
Which points?
Answer: All points the same distance r from a given point, the center.

We like to abstract things in math.
We can define a *circle* as a *point* and a *number*.
The **point** is the **center**, and the **number** is the **radius**.

Let the center be at the origin.
$(h,k) = (0,0)$.
The equation for a circle is then
$$x^2 + y^2 = r^2.$$

6.2.1   Ellipses

We can generalize this to

$Ax^2 + By^2 = C$. This is an ellipse.

If $A \neq 0$, or $B \neq 0$, this equation has two parameters, along with the center (a point).

Contrast this with a circle that has one parameter, the radius. It also has the given center point. An ellipse has two radii along with the center point. Instead of

$Ax^2 + By^2 = C$, we will discuss the simpler equation
$Ax + By = C$ in the next section.

6.2.2   Student errors

Here is another thing that caused confusion and errors.

Find the equation of a circle center at $(0, 0)$ and a point on the circumference at $(5, 2)$.

The answer is that the equation of a circle center at the origin and radius r is

$x^2 + y^2 = r^2$. We set $x = 5$ and $y = 2$, and get $r^2 = 29$.
The equation is
$x^2 + y^2 = 29$.

Some students wrote the equation as $5^2 + 2^2 = 29$.

They confused variables like x and y with numbers like 5 and 2. An equation of a circle involves variables, which are the coordinates in the x-y plane.

Another error was saying the distance between the points $(0,0)$ and $(5,2)$ is 29, when it is the square root of 29.

6.3   Lines and functions

Instead of the quadratic equation of an ellipse, we will discuss the same equation as a linear equation:

$Ax + By = C$.

This is a line. This is the general equation of a line.

To understand this better, let us contrast this with the equation

where x and y are squared: $Ax^2 + Bxy + Cy^2 = D$.

This is a lot more complicated, as it is not linear.

If $B = 0$ and $A = C$, this is a circle, and D is the radius squared.

Let us return to the linear equation $Ax + By = C$.

Note that if $A \neq 0$ we can divide every term by A.

Likewise if $B \neq 0$ we can divide every term by B.

This means that a general line is specified by 2 parameters.

If $B = 0$, we get a vertical line.

Since y does not appear, y can be anything.

If $B \neq 0$, we divide by B and get

$$y = \frac{-A}{B} x + \frac{C}{B}.$$

By convention, we write the constants as

$y = mx + b$, where $m = \dfrac{-A}{B}$ and $b = \dfrac{C}{B}$.

The parameter m is the slope, and b is the value of y where $x = 0$.

This is called the **slope - intercept form** of the equation of a line.

This is a function: $y(x)$.

A function is an expression: given input x we get one output.

The possible values of the input is called the **domain**.
The possible values of the value of the function is called the **range.**

"Home, home on the range…"

Give something in the home *(x)*, we get something in the range *(f)*.

Some teachers discuss lines starting with the slope-intercept equation. The logical starting point is the general linear equation, with the slope-intercept form a special case for lines that are not vertical.

Here is another way to view functions.
A **relation** is a set of ordered pairs.
A **function** is a relation where there is one *y* for any *x*.

We need *two things* to define a non-vertical line. E.g., *m* and *b*, slope and an intercept, or slope and a point. Note that there are two intercepts, the x-intercept and the y-intercept. For vertical lines, we also have two things: the *x* coordinate, and the fact that the line is vertical.

Let us now consider two lines:
$y = m_1 x + b_1$ and $y = m_2 x + b_2$.
If the lines meet, the intersection, the point (x, y) is the solution of these two equations.

If $m_1 = m_2$, there is no solution to the two equations. This means the lines do not meet. There is no common point. The lines are parallel, which they should be as they have the same slope.
If $m_1 m_2 = -1$, the lines are perpendicular.

The intersection of the perpendicular lines is the solution of the two equations. The proof of this is in the next section, entitled "Slope theorem".

6.3.1   Solutions of linear equations

If two lines intersect, then there exists a solution of the two linear equations for each line.

You can solve linear equations by a combination of the following:

1. *Substitution*. Solve one equation for, say, y, replace it in the other equation, and solve for x. Now go back to the first and solve for y.

2. *Elimination*. Multiply an equation by a number to simplify. Add equations to eliminate one variable.

For three equations in three variables, you may need to use both of these methods. First look at two equations, solve with the third variable being left as a variable. Then look at the third equation.

6.3.1.1   What does solving an equation mean?

We tell our students how to solve equations, without verifying that they indeed know what we mean by solving an equation.

The first step is asking what is an equation. The answer is something with an equal sign.

Problem solving is a game. I tell you things. I may tell you two things. These are two statements, that is, two equations involving variables x and y. You then tell me what theses two variables are. This is what we mean by solving equations.

6.3.2 Slope theorem

Theorem: If the product of the slopes of two lines $= -1$, the lines are perpendicular.

Let one line have slope m_1 and the other have slope m_2.

We will prove that $m_1 m_2 = -1$.

Let us draw the first line through the origin of a coordinate system. Show this by drawing a right triangle with length 1 and height m. The slope of the hypotenuse is m. The angle between the hypotenuse and the horizontal x-axis is α. The angle between the hypotenuse and the vertical y-axis is β. We note that

$$\alpha + \beta = 90°$$

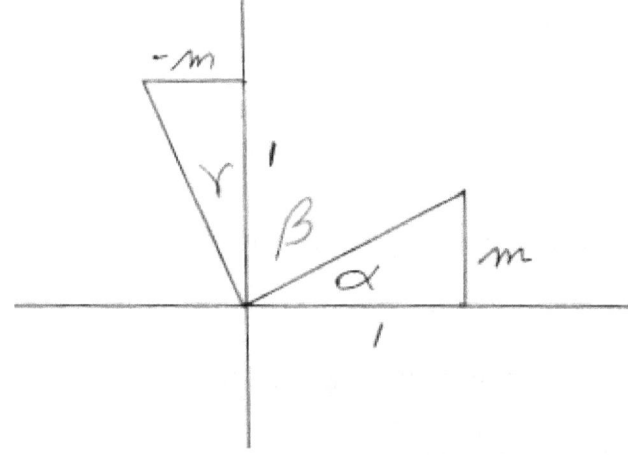

We draw another triangle, height 1 and length $-m$.

The slope of the hypotenuse is $\dfrac{1}{-m}$.

That is, $m_1 = m$ and $m_2 = -\dfrac{1}{m}$.

The two triangles are congruent, as they are right triangles with two adjacent sides equal. Therefore $\gamma = \alpha$, and so $\gamma + \beta = 90°$.

This proves that the lines are perpendicular, and that the product of the slopes is negative 1.

6.4   Translations

I look at a suspect 50 ft away, and call another guard 300 ft. away.
From the other guard's point of view, he looks back at the suspect.
This idea of translation can be extended to our social lives. When
dealing with people, we need to look at things from the other person's
point of view. Very young children are incapable of this. This is a skill
we learn as we mature. We need to consciously translate ourselves to
the other person, asking ourselves how things look from their point of
view.

Let's look at the mathematics.

Translation between coordinate systems (x, y) and (x', y').

E.g, I am at the origin of x,y. I am a security guard. My
coordinate location is $(0,0)$. Another guard, Joe, is at (h, k).
I see a suspect at a point (x, y).
Joe will see the suspect at (x', y'). This point is found by a linear
transformation: $x' = x - h, \; y' = y - k.$

E.g., I see the suspect 50 ft ahead. Let Joe be 300 ft ahead.
Then $h = 300, \; k = 0$.
Joe sees the suspect at $x' = -250$.

We do the same thing with circles.
The equation for a circle at the origin is $x^2 + y^2 = r^2$.
If I see the center of the circle at (h, k), then Joe sees the circle at
his origin.
Joe's equation for the circle is $x'^2 + y'^2 = r^2$.
The equation for the circle that I see is $(x - h)^2 + (y - k)^2 = r^2$.

Here is an explanation for translations:
http://www.math10.com/en/geometry/analytic-
geometry/geometry1/coordinates-
transformation.html

We can do the same thing with lines. Draw a non-vertical line through the origin. The equation is $y = mx$. Joe draws a line through his origin: $y' = mx'$.

I see Joe's line using the linear transformation, and get a line with the usual form $y = mx + b,$ where this m is not the same m that we started with.

Let us apply the transformation from $y' = mx'$ using $x' = x - h,\ y' = y - k.$ We get
$y - k = m(x - h).$

Simplify:
$y = mx + k - mh.$

This has the standard form if $b = k = mh,$ resulting in $y = mx + b.$
The parameter b is the y-*intercept,* the value of y when x is 0.

What is interesting is the we start with the simplest thing possible and move forward. Mathematicians like to start with simple things. For lines, we start with a diagonal line through the origin, apply a transformation, and get the usual form of a line with a non-zero y-intercept.

The same is true for circles. We start with the simplest circle, a circle with the center at the origin, and apply a transformation to find the equation of a circle with the center anywhere.

6.5   Exponential functions

Lines are linear functions. By this, we mean an equation for a circle has y raised to the power 1. $Ax + By = C$, or $y = mx + b$.

Circles are quadratic functions: $y^2 = r^2 - x^2$. This means that y is raised to the power 2.

An exponential function $y(x)$ has containing something raised to the power x: $y = b^x$.

Usually we take e, Euler's constant, instead of any number b. The reason is that if we draw the graph $y = Ae^x$, where A is a constant, we will find that at any point, that is, for any value of x, the slope of a line tangent to this graph will equal the value of y. This we prove in the next section.

Let us look at some examples where the change is proportional to the value. One is money in the bank earning interest. The change in the amount of money is equal to the interest, if the interest is done continuously. This is not so for interest at certain periods. If the interest is calculated monthly, then each month the amount jumps by the interest payments. For continuous interest, the amount of money earned as interest depends upon the amount in the bank. If you have twice the amount of money in the bank, you will get twice the amount of interest.

The formula for continuous interest is this. Let A be the amount after t years, with **r** the rate per year, and P the principal (initial deposit). The formula is $A = Pe^{rt}$.

The same is true for radioactive decay. For example, tritium has a half-life of 12.3 years. This means if you have one pound of tritium, after 12.3 years you will have a half a pound.

6.5.1 Change in exponential function

If we multiply 1 by 1, we get 1, even if we do this many, many times. However, what happens if we take 1 plus a small number, and multiply it by itself many times? It turns out we get a constant, called Euler's number e.

This has the unique property that the change of e^x is e^x.

The change is equal to the value of the function. No other function has this special unique property. For example, for a line, the change is constant.

We can prove this using calculus. Here is a proof that does not use calculus.

The formal definition is $e = \left(1 + \dfrac{1}{n}\right)^n$, with $n \to \infty$.

Let $y = e^x$.

The change in y, which is Δy, is $e^{x+\Delta x} - e^x$.

Recall the change from a to b is $\dfrac{b-a}{a}$ or $\dfrac{\Delta a}{a}$.

We will prove that $\dfrac{\Delta y}{\Delta x} = e^x$.

Proof :

$\Delta y = e^{x+\Delta x} - e^x$.

We use the definition of e :

$e = (1 + \dfrac{1}{n})^n$, where n is very large.

Our equation becomes, noting that $e^{x+\Delta x} = e^x e^{\Delta x}$,

$\Delta y = e^{x+\Delta x} - e^x = e^x(e^{\Delta x} - 1)$

$$\Delta y = e^x \left(\left(\left(1 + \frac{1}{n}\right)^n \right)^{\Delta x} - 1 \right)$$

$$= e^x \left((1 + \frac{1}{n})^{n\Delta x} - 1 \right),$$

as the exponent of an exponent is
the product of the exponents.

Since Δx is very small, we will write $\Delta x = \dfrac{1}{n}$.

$\Delta y = e^x \left((1 + \Delta x)^{\Delta x / \Delta x} - 1 \right).$

$= e^x \Delta x$

Or

$\dfrac{\Delta y}{\Delta x} = e^x$

The change in y is equal to y. For this function, the change of the function is the function itself. In the general case, the change in the function is proportional to the function.

Let $y = Ae^{Bx}$. The change in y is By.

6.5.2   Compound interest

Having discussed continuous interest, let us briefly discuss compound interest, so that you can clearly compare them.

Let A be the amount of money in the bank after t years, and P the initial amount. The bank gives interest at the rate of r. This is always expressed as percent per year. The bank compounds the interest n times a year. If $n=12$, this means that 12 times a year the bank adds interest. The rate per period is r/n.

First, let us consider simple interest.
$$I = \operatorname{Pr}t$$
$$A = P + I = P(1 + rt).$$

For compound interest, for example, after one month, the amount would be

$$A = P\left(1 + \frac{r}{12}\right).$$

After 2 months, we multiply the new amount by $1 + \frac{r}{12}$:

$$A = P\left(1 + \frac{r}{12}\right)^2.$$

After 1 year, the amount would be $A = P\left(1 + \frac{r}{12}\right)^{12}$.

In general, if something compounded n times a year, the amount after t years is

$$A = P\left(1 + \frac{r}{n}\right)^{nt}.$$

As n goes to infinity, we see, using the definition of e, that the expression for A is

$$A = Pe^{rt}.$$

6.5.3   Scientific notation

One trillion is 1 with 12 zeros. That is, one trillion $= 10^{12}$. Likewise, one billion has 9 zeros, a million has 6 zeros, and a thousand has 3 zeros.

The nearest star, Alpha Centauri, is 25 trillion miles from earth. Until recently, the number trillion was reserved for astronomical distances. Now, unfortunately, it also is used for U.S. debt.

6.5.4   Some student errors

Simplify $6p^{-7}$.

Some students wrote $p^{\frac{6}{7}}$. Can you find the error?

What is important is not only to get the correct answer, but to understand errors, your wrong ways of thinking.

Recall the proof: $3^2 \times 3^{-2} = 3^0$, as we add exponents.

Since $3^0 = 1$, we can divide both sides of this equation by 3^2:

$$3^{-2} = \frac{1}{3^2} = \frac{1}{9}.$$

Students know that negative exponents means the number goes in the denominator. The error is putting it in the denominator of the exponent!

$3^{-2} \neq 3^{\frac{1}{2}} = \sqrt{3}$. On the other hand, $3^{-2} = \frac{1}{9}$.

Here is another mistake:

$$9\left(4p^3\right)\left(6p^{-7}\right) = 36p^3\left(6p^{-7}\right) = 36p^3 \times \frac{1}{\left(6p^7\right)} = 6p^{-4}$$

The correct answer is $9 \cdot 4 \cdot 6 \cdot p^{-4}$.

Speaking of mistakes, here is an error involving ratios.

The country saved $36 billion, while the debt is $14 trillion. The ratio of saving to debt is 14,000/36. The error is that when we say ratio, we mean that the saving is in the numerator.

The answer is $\dfrac{36}{14,000}$, which is about 2.8×10^{-3}.

To change to percent, multiply by 100, and get 0.28%.

Another confusing thing:

Sales were $285 billion, and profits were $19 billion. The government increased taxes by $4 billion, and now the profit went down to $15 billion. Find the amount of decrease of the ratio of profits to sales. Assuming the sales figures did not change, even though they would change in reality.

The decrease is (new ratio – old ratio)/old ratio. This is

$$\frac{\left(\dfrac{15}{285} - \dfrac{19}{285} \right)}{\left(\dfrac{19}{285} \right)}.$$

7. Counting

How to count systematically.

In mathematics, we do things systematically. We do not just *do* things.
Consider brushing your teeth. Whoosh, whoosh, whoosh! No! You
must brush each tooth, brushing in sequence, being to brush the front,
bottom (or top) and back. If you are careless, you may develop
expensive dental problems. Put down 10 pieces of chalk, and ask a
child to count them. Children just move their fingers randomly over the
chalk, and often do not get the correct answer. You must start with
the left piece, move your finger to the next piece, to be sure you
counted correctly.

This chapter deals with systematic counting.

7.1 NJ license plates

As an example of the need for systematic counting, consider the
problem counting license plates. In 2009 New Jersey realized that
they have to change the sequence of letters and numbers, for the
present sequence would not permit enough plates for the anticipated
drivers.

Here is what was written on their site:

http://www.newjerseynewsroom.com/state/nj-license-plates-will-change-next-year

"Up until now, the so-called standard issue plates have started with a
sequence of three letters, followed by two numbers and a letter. But
the state Motor Vehicle Commission has just started issuing plates
that start with ZZZ, so a change is in order.

"When the final plate in the series, ZZZ-99Z, is issued, sometime next year, the sequence of letters and numbers will be reversed, the Associated Press reports.

"The change will yield 25 million new combinations. Officials said the new series, which will begin with A10-AAA and go through Z99-ZZZ, will likely last about 20 years."

How did they know the change would yield 25 million new combinations? They did not get state workers writing down all possible combinations and counting them. Instead, mathematicians have developed principles and the application of these principles permit this calculation to be done easily.

7.2 Fundamental Counting Principle

One principle is the *Fundamental Counting Principle.* If a task consists of 2 parts, and the first can be done in n ways, and the second in m ways, the number of ways of doing the entire task is the product $n \cdot m$.

This is true only if a certain criterion is met, the *uniformity criterion.* This is that if the number of ways of choosing the second part is the same no matter what choices were made for the first part.

Of course, instead of 2 parts, we can use a variable, and talk about k parts. If the number
of ways of doing the *ith* part is n_i, the number of ways of doing the entire task is the product of the ways of doing each part:

$$\prod_{i=1}^{k} n_i.$$

To better understand the capital pi, imagine that $k = 2$. In this case the number of ways is $n_1 n_2$.

Here the uniformity criterion means the number of ways of doing the *ith* part does not matter what choices were made for the previous parts.

Note that the uniformity criterion deals with the number of ways, not the the choices. The choices of doing the second part actually depend on the choices of doing the first part. The uniformity criterion deals with counting, the number of ways, not the choices.

A source of difficulty is reading. You need to read phrases (or single words, if it is a big word), and stop reading, and think. As an example, consider the words *"uniformity criterion"*. Think of the meaning of *criterion*. I asked a student her criterion for choosing boys. Then she understood what criterion meant.

Here is another example. Consider the phrase *"multipart task"*. Visualize an example. E.g., cleaning up after dinner. Clean the table, then sweep. This is a two-part task. How many ways to use two digits to get a number? First, choose the first digit, then choose the second. This is also a two-part task.

7.3 Ways of arranging n distinct objects

How many ways can we arrange n distinct objects?

There are n ways to choose the first object.
There are $n-1$ ways to choose the second object,
as we already chose the first.
The uniformity criterion is satisfied here, as the number of ways
of choosing the second object does not matter what we chose for
the first. We keep on multiplying, until we get one object left, and
then there is one way to choose it.

We can write this mathematically as the number of ways of arranging
n objects is

$$n(n-1)(n-2)\cdots 1 = \prod_{i=0}^{n-1}(n-i)$$

This is the definition of factorial, $n!$, for $n > 0$.
For $n = 0$, we note that since there is only one way to arrange 0
objects, we define $0! = 1$.

Note that the ! factorial operator is a function, even though it comes
after the number. To be consistent with writing functions, one should
write it as !n, like in Spanish. However, it is written after the number.

7.3.1 Look-alikes

If the n objects have a subset n_1 that has objects that look alike, we divide by the number of ways of arranging the n_1 objects, which is $n_1!$. If there is another subset n_2 that has objects that look alike, we divide by $n_2!$ also.

In general, we divide n! by the product of the number of ways of arranging the subset of objects whose arrangements we do not care. The number of distinguishable objects, where n_i subsets look alike, is

$$\frac{n!}{\prod\limits_{i=1}^{k} n_i!}.$$

If there are two subsets that look alike,

the number of subsets is $\dfrac{n!}{n_1! \times n_2!}$.

7.3.2 Permutations

By the permutations of the letters *abc* we mean all of their possible arrangements, where order matters $(ab \neq ba)$:

abc
acb
bac
bca
cab
cba

There are 6 permutations of 3 different things. As the number of things (letters) increases, their permutations grow astronomically. For example, if 12 different things are permuted, then the number of their permutations is 479,001,600.

Now, this enormous number was *not found by counting them*. It is derived theoretically from the *Fundamental Principle of Counting*, which says there are $n!$ ways to arrange n objects.

To find the number of ways of choosing r distinct objects from n objects, we note that we do not care how the $n - r$ remaining objects are arranged. We care only about the order of the r selected objects. We divide $n!$, the number of ways of arranging n objects, by $(n - r)!$:

$\dfrac{n!}{(n-r)!}$. We write this as $_nP_r$.

If $r = n$, we divide by $(n - n)! = 1$.

Therefore we get $_nP_n = n!$.

The number of ways of choosing n objects out of n is $n!$, as we expect.

The text by Miller discusses this another way. To choose 2 objects out of n, we note that there are n ways to choose the first, and $n - 1$ ways to choose the second. Multiply and divide by $(n - 2)!$ to get the expression above. Personally I find this confusing for a number of reasons. For example, why do we have to multiply and divide? The formula says that if we pick r objects, we do not care about how the remaining $n - r$ objects are arranged.

7.3.3   Combinations

Note that we divided by the number of ways of arranging things we do not care about. Suppose we do not care about the number of ways of arranging the r objects that we chose.
We then divide by $r!$.

This is called **combinations:** $\dfrac{n!}{(n-r)!r!}$.

Example: Suppose n is 3 and r is 2.
Given 3 digits, we want to write all possible 2-digit numbers, where no number repeats. Let the 3 digits be $\{1,2,3\}$.

There will be $\dfrac{3!}{(3-2)!} = 6$ numbers:

12
13
21
23
31
32

This is the list of permutations of the numbers. If we asked for the combinations, we would also divide by $r!$, where $r = 2$. Our list is now

12
13
23

7.3.4   Summary of permutations and combinations

In summary, the number of ways of choosing r objects from n, called **permutations**, is

$$n \Pr = \frac{n!}{(n-r)!}$$

For **combinations,** we do not care about the order of the r objects, and so we divide by the number of ways of arranging r objects, which is $r!$. We have

$$nCr = \frac{nPr}{r!}$$

$$= \frac{n!}{(n-r)!r!}$$

What is the sum of all the combinations of n things?

$$_nC_0 + {}_nC_1 + {}_nC_2 + \ldots + {}_nC_n = 2^n$$

The combination is the number of subsets, as order does not matter with sets. The above is the total number of subsets.

7.4   Sometimes need to change the question

Sometimes we have to change the question to count the ways.

How many ways can 4 golf clubs be given to 3 sons?
Instead of looking at each son counting how many ways a son can get a club, look at each club, and see how many sons can get it.

7.5 Cards

Many problems involve cards, such as asking how many ways can a black card be chosen. A student said he did not know how cards are arranged in a deck. Here is the answer.

There are two colors, black and red. There are 4 suites: spades and clubs are black, and hearts and diamonds are red. Each suite has 13 cards. The first one is called the Ace, and it is card number 1. The numbers go from 1 to 10. This is followed by 3 picture cards: Jack, King, and Queen. The total number of cards is 13×4, or 52.

Here is a picture of the 52 cards.

Example set of 52 poker playing cards

Suit	Ace	2	3	4	5	6	7	8	9	10	Jack	Queen	King
Spades													
Hearts													
Diamonds													
Clubs													

8. Six principles of counting

There are six principles for counting:

1. *Repeats allowed.* This is the fundamental principle of counting. We multiply number of ways of doing each part.

2. *Distinguishable* arrangements containing look-alikes.

3. *Repeats not allowed, order matters.* This is permutations.
The number of ways is $n\text{Pr}$.

4. *Repeats not allowed, order does not matter.*
The number of ways is $n\text{Cr}$.

5. *NOT*. We want to count all the arrangement except something.
E.g., flip three coins. How many ways can we get anything except all tails? We use set theory, where A is the set of desired arrangements. This is equal to the set of all possible arrangements, which is the universal set U, minus the set of set of things not desired. Since the number of desired arrangements plus the set of undesired arrangements is the set of all arrangements we have
$n(A) = n(U) - n(A')$.

Note that here we subtract, and not multiply as we do for the other cases.

This is also called the *complements* principle.

6. *OR*. We use set theory, cardinality of union of sets. Let A be the set of arrangements satisfying one condition and B be the set of arrangements satisfying another condition. The number of arrangements satisfying one or the other condition is
$n(A \cup B) = n(A) + n(B) - n(A \cap B)$.

9. Binomial theorem

Consider the product $(x+y)^2$. We can write this as $(x+y) \cdot (x+y)$. This is a product of two factors.

When we multiply, we multiply the first x's. Then we multiply the first x times the y. We then multiply the y times the x, and finally multiply the y's. We get $x^2 + xy + yx + y^2$. Since numbers are commutative, we can write this as $x^2 + 2xy + y^2$. What we have done is first to choose an x from the first factor and an x from the second factor. There is only one way to do this. We then choose an x and an y. There are two ways of doing this. We can view this using the idea of combinations. How many ways can we choose a y from 2 objects? The number of ways of choosing 0 y's is $_2C_0 = 1$. The number of ways of choosing 1 y is $_2C_1 = 2$. The number of ways of choosing 2 y's is $_2C_2 = 1$.

We can generalize this to looking at $(x+y)^n$. We get terms like $_nC_r x^{n-r} y^r$.

This is the binomial theorem.

Usually, instead of C, we write $\binom{n}{r}$. This is the same thing.

We want to expand $(x+y)^n$. First, let us look at $(x+y)^2$.

$$x+y$$
$$\underline{x+y}$$
$$x^2+xy$$
$$\underline{yx+y^2}$$
$$x^2+2xy+y^2$$

Now let us look at $(x+y)^4$. To do this,

we take one term (an x or a y) from each of the four factors.

E.g., let us look at one of the terms: x^2y^2.

We have to pick two x's out of 4.

How many ways can we pick two x's? There are no repetitions.

We say that order does not matter, as $xy=yx$.

This means we have a combination, $_2C_2$.

Question:

Why does this expansion use combinations and not permutations?

The answer is that we start from the left and move in sequence.

For example, we can take an x from the first and third factors.

Remember that there are four factors $(x+y)$.

If the order is important, that means choosing

from the third and first is different from choosing

from the first and third, we use permutations.

10. Probability and statistics

We can define probability by looking at sets. If E is the set of desired events and S the sample space, we define the probability of the event E as the ratio of the cardinalities of the sets.

Formally, we write $P(E) = \dfrac{n(E)}{n(S)}$.

For example, let us look at flipping a single coin.

We have the probability of a set and the probability of a number. For example, flip two coins.

$$P(\{h,t\}) = \frac{1}{4}.$$

In this case, the set $E = \{h,t\}$.

Let us find the probability of one head. Let the number of heads be x. We have $p(x) = \dfrac{1}{2}$, for $x = 1$. Again, we have the probability of a set, $P(E)$, and the probability of a number $p(x)$.

10.1   Theoretical and empirical probability

Flip a coin. The theoretical probability of getting a head is ½, as we discussed.

The empirical probability is found by actually doing it. We ask the class for each student to flip a coin 10 times and count the number of heads. If 15 students did this, the cardinality of the sample space is 150. We count the number of heads. If each student got 5 heads, the event probability would be 75, and the empirical probability would be ½. Actually, some student got 5 heads, some 7 heads, and some 3 heads. There were 33 heads out of 70 tosses, close to the theoretical value of 35. We imagine that if there were 700 tosses, the value would likely be closer to 350 heads.

See
http://www.zweigmedia.com/ThirdEdSite/tutorialsf2/frames6_3.html.

The larger the sample space, the closer the empirical probability is to the theoretical probability. This is called the *Law of Large Numbers*. The proof of this is beyond the level of the course. One way of saying this is this is the empirical probability is an approximation to the theoretical probability. When we speak of an approximate measurement, we mean that there is a difference (an error) between the approximate measurement and the theoretical measurement. The Law of Large Numbers says that this error becomes smaller the larger the number of trials.

Here is a gif file showing how random blue or red colors are generated, demonstrating the Law of Large Numbers:
http://en.wikipedia.org/wiki/File:Lawoflargenumbersanimation.gif

We worked on a survey problem above. We showed that out of 240 families who had dogs, cats, and parakeets, 19 had a parakeet only. In this problem the probability of finding a family who had only a parakeet is 19/240.

10.2 Odds

Probability is a number, e.g., $\frac{8}{13}$, which is about 0.6.
Probability is always ≥ 0 and ≤ 1.

Odds are two numbers, such as 8 to 5.
Two numbers are not the same mathematical entity as one number.
Odds using two numbers is another way to express probability (probability is one number).

10.3 NOT, OR, AND

The *NOT* **rule** is very simple, very similar to the NOT rule for counting discussed above.

P(not E) = 1 - P(E).

This follows from the definition of the complement of a set: $S = A + A'$. Take the cardinalities of both sides, and divide by $n(S)$. The sample space S is the universe U in Venn diagrams.

This is called the *complements* principle.

The *OR* **rule** is also similar to the OR rule for counting discussed above.

To find *P(A OR B)*, we use $n(A \cup B) = n(A) + n(B) - n(A \cap B)$.

Divide by $n(S)$. We note that *P(A OR B)* $= \dfrac{n(A \cup B)}{n(S)}$.

Therefore:

$P(A \cup B) = P(A) + P(B) - P(A \cap B).$

If the intersection of A and B is zero, it means that both cannot happen simultaneously. That is, the events are mutually exclusive. The rule is this: **The probability of either of two mutually exclusive events happening is the sum of the probabilities of each event**.

Problem: If license numbers consist of three letters followed by three digits, how many different licenses could be created having at least one letter or digit repeated?

An example of the OR rule is finding the probability of picking one card from a deck, where the card is either red or a spade. In this case, the probability is $P(\text{red}) + P(\text{spade})$.

The AND rule is based upon the fundamental counting principle. If A and B are independent, that is, event A is independent of event B, then the probability of both A and B is the product $P(A) \times P(B)$. An example is two people picking cards. *A* is the probability one person picked a spade from a 52 card deck, and *B* is the probability the other person picked a red from a 52 card deck. The probability of this happening is $P(\text{red}) \times P(\text{spade})$.

Mutually exclusive events are such that if one happens the other cannot happen. If we pick a red card, then we did not pick a spade. Independent events are events that have nothing to do with each other, such as choosing cards with replacement. Since we put the card back, the choice of the first card does not affect the choice of the second card.

10.3.1 Example

Find the probability of taking two black cards from the black cards from a 52 card deck, where none of the cards are aces.

One way to do this is first to note that there are 26 ways to pick the first card, and 25 ways to pick the second card. Since we do not care about the order, we divide by 2. Since we do not want aces, we actually have 24 ways to pick the first card and 23 ways to pick the second card. The probability is then

$$\frac{24 \cdot 23}{2 \cdot 52 \cdot 51}.$$

Suppose we wanted to arrive at this conclusion by following the principles outlined above. Although this is a much longer way, it is instructive.

The probability of picking up a black card is $\dfrac{26}{52}$.

The probability of again picking up a black card, where we do not care for the order of the cards in our hand, is $\dfrac{26 \cdot 25}{52 \cdot 51 \cdot 2}$.

This is a combination. We can also find this using the expression for combinations. The probability of picking up one black card from the black cards is:

$$\frac{26!}{(26-1)! \, 1! \cdot 52} = \frac{26}{52}.$$

We multiply by the probability of also picking up another black card and get $\dfrac{26}{52} \times \dfrac{25}{51 \cdot 2}$.

The probabity of not picking up an ace is equal to 1 minus the probability of picking up one ace: $\left(1 - \dfrac{2}{26}\right)$.

The probability of not picking up another ace is $\left(1 - \dfrac{2}{25}\right)$.

We multiply all the factors:

$$\frac{26}{52} \times \frac{25}{51 \cdot 2} \times \left(1 - \frac{2}{26}\right) \times \left(1 - \frac{2}{25}\right) =$$

$$\frac{26}{52} \times \frac{25}{51 \cdot 2} \times \frac{24}{26} \times \frac{23}{25} =$$

$$\frac{24}{52} \times \frac{23}{51 \cdot 2}.$$

10.4   Example problem

A pet store has 4 poodles, 2 terriers, and one retriever. If Sandra and Joe each select one puppy at random, and each may both select the same one, find the probability that Sandra selects a retriever and Joe a terrier.

Students made errors in reading the problem. We need to focus on correct reading!

The problem was Sandra picks a retriever and Joe a terrier.

The error was reading it like this: Find the probability that Sandra picks a retriever and find the probability that Joe picks a terrier. There are two answers to this question. There is one answer to the original question.

To find the answer, we find the probability that Sandra picks a retriever and multiply it by the probability that Joe picks a terrier. We multiply because of the AND.

The answer is $\dfrac{2}{7} \times \dfrac{1}{7}$.

10.5 Percentages

Percentages are simply numbers divided by 100.

10.5.1 Blonde girls

I overheard a student asking, "Why do we have to learn percentages?" It is important to be aware of the students, to listen to them, and to respond. When we respond to their questions and displays of interest, we get better educational results.

To deal with this, I started by asking the very simple question,

"What percentage of the students are boys?"

They replied, "50%".

"This seems right," I responded.

I then asked, "What percentage of girls are blonde?"

One girl popped up and said, "Do you mean natural blonde or all blondes?"

"Let's say natural blondes," I replied. "This means if we have 100 girls, how many are natural blonde?"

The important thing about statistics is to use whole numbers, i.e., frequencies, not decimal fractions, in order to make it more meaningful.

"About 15 girls."

"Okay. This means that 15% of the girls are natural blondes. Now how many girls are blonde, natural or not?"

"About 40%."

"Good. Now how many blonde girls are natural blonde? If we have 100 girls, 15 are natural blonde, and 40 are blonde. The number of natural blondes out of all blondes is *15/40*.

They started figuring. I said to make it faster, as the bell was going to ring, try to double numerator and denominator. We get 30/80. If we make numerator and denominator a little larger, so that the denominator is 100, the numerator will also be larger. Say 40. It means that 40% of blonde girls are natural blonde. This means that 60% of the blonde girls color their hair.

We now see the importance of percentages. We now discovered how many girls are not natural blonde.

I then asked the girl who popped the question of natural or not whether she is natural. She said she was. Well, she is special, one of the few blonde girls who are natural. Now she feels good. Now the students got the point of the need for percentages.

When we talk about something meaningful and important to them, such as hair color, their interest and attention increases.

Again, when we talk about percentages, such as 40% of the blonde girls are natural blonde, we do not give them the picture that 0.40 blonde girls are natural, but that 40 girls out of 100 blonde girls are natural. It is hard to think of 0.40 girls, as opposed to 40 girls.

Adults also get confused with percentages. Frequencies, which are whole numbers, are much easier to comprehend than decimal fractions. If the adult finds a concept confusing, the student certainly will be confused. We should try to minimize confusion by using whole numbers.

As a general rule, we have to examine ourselves, and if we find something the least bit confusing, we must think it through carefully and plan how to present it.

10.5.2 When a doctor suggests surgery

When a doctor suggests surgery, ask two questions:

1. Out of 100 people, how many do well? Note: Do not ask for the probability, which is a fraction $<= 1$, but say out of 100 people. If you did this procedure on 100 people, how many would do well?

2. For those who did not do well, what was their condition? E.g., if 95 out of 100 would do well, what is the situation of the other 5?

The important thing to realize that probabilities expressed as percents do not mean as much to people as numbers of people out of 100. I read this in a book a long time ago.

If the doctor says that he did not do this procedure on 100 people, see another doctor.

The above discussion assumes the doctor is telling you the truth. Unfortunately, doctors sometimes lie. Mathematics is based upon arbitrary assumptions. If the assumption (telling the truth) is false, so is the conclusion.

10.5.3   Example from class

At the start of the day, we noticed that 5 students were absent. I asked the girls next to me what percentage of the class was absent. They felt the question was too hard to answer. This a further example of making the students think of the material by using real life situations.

10.5.4   Example from news

Drug studies with Vioxx found that 0.75% people on a placebo got heart attacks and strokes, vs. 1.5% who used Vioxx. The article stated that Vioxx caused a doubling of the risk, as 1.5 / 0.75 is 2. What it is actually saying is that out of 100 people, 0.75 more would get a heart attack or stroke. Saying 0.75 more would get a heart attack or a stroke does not sound as bad as saying the risk is double, even though both statements are correct!

The statement given in the news media is actually misleading. If we have to make a decision whether or not to take a certain drug, we do not care if the risk is double than the placebo. We merely care that the risk is larger than the margin of error. What we care about is what would happen to 100 people who take the drug. For example, suppose the risk with the placebo were 0.25%, and with the drug 1.00%. Here the risk is four times as large as the risk with the placebo. However, out of 100 people, 0.75 more would suffer. This is exactly the same as the above example with the double risk!

This can be given to illustrate how the news media present information which although may be factually correct, does not convey the proper informative content.

Another example is with condoms. Young people are urged to use "protection" when having sex, which means using condoms. Supposedly, condoms are 99% effective against HIV. What does this mean? Does this mean that if a man had sex with 100 women who have HIV, he would get HIV? No, this is not what it means. When I say this to students, they break out in laughter! It means that if 100 couples had sex, where one person in each couple had HIV, and they used condoms, one couple would get HIV. If we say condoms are 99.9% effective, it means that if 1000 couples had sex with condoms, one couple would get HIV. This is on the average, for it is possible that one sex act with a condom can cause HIV.

10.6   Conditional probability, AND multipilication

Conditional probability:
Instead of dividing by $n(S)$, we divide by $n(B)$, where B is the space satisfying the condition.

Let A and B be sets. **The conditional probability $P(A \mid B)$, is the probability of A given B.**

Example: pick a jack on the condition you pick a black card.
A is the set of black jacks.
B is the set of black cards.

Another example:
$S = \{1, 2, \cdots, 10\}$.

A = set of multiples of 3 from S
B = set of odd elements from S
$P(A|B)$ is the probability of choosing a multiple of 3
from S where you chose an odd element.

$$P(A \mid B) = \frac{n(A \cap B)}{n(B)},$$

as this is the number of both A and B divided by the number of B.

Divide numerator and denominator by $n(S)$. We get

$$P(A \mid B) = \frac{P(A \cap B)}{P(B)}.$$

Likewise, since A and B could be anything:

$$P(B \mid A) = \frac{P(A \cap B)}{P(A)}.$$

In English words, this says that the conditional probability of B given A is the probability of both A and B divided by the probability of A.

If A and B are independent, then
$P(B \mid A) = P(B)$, as the probability does not depend on A.
We get $P(A \cap B) = P(A)P(B)$ if they are independent.
This is the **multiplication** rule.

Summary for independent probabilities:
AND — multiply probabilities
OR — add probabilities

Note again that these rules for AND and OR are different if the probabilities are not independent.

10.6.1 Bayes' theorem

Bayes' rule is a relationship between $P(A \mid B)$ and $P(B \mid A)$.
Let B_1 and B_2 be mutually exclusive and exhausive events,
such that $P(B_1) + P(B_2) = 1$.
Let A be an observed event.
We write the conditional probability

$$P(B_1 \mid A) = \frac{P(B_1 \cap A)}{P(A)}.$$

Multiply and divide by $P(B_1)$. We get

$$P(B_1 \mid A) = \frac{P(B_1 \cap A)}{P(A)} \frac{P(B_1)}{P(B_1)}$$

$$= P(A \mid B_1) \frac{P(B_1)}{P(A)}.$$

We note that

$$P(A) = P(A \cap B_1) + P(A \cap B_2).$$

Therefore

$$P(A) = P(A \mid B_1)P(B_1) + P(A \mid B_2)P(B_2)$$

$$P(B_1 \mid A) = \frac{P(A \mid B_1)P(B_1)}{P(A \mid B_1)P(B_1) + P(A \mid B_2)P(B_2)}.$$

This is Bayes' rule.

Here is an example.
Given the following facts:

Annual family income	% married couples	% both employed
$150,000 or over	4	65
$100,000 - $149,999	10	73

This means that 4% of married couples have income $150,000 or over. Out of these people, 65% of both spouses are employed.

Here are some questions we can ask:

What is the probability of a couple selected at random with family incomes $100,000 or over has both spouses employed?

Secondly, for a randomly selected couple having incomes $100,000 or over and having with two incomes, what is the probability that the family income is $150,000 or over?

Let A be the set of couples with income $150,000 and over, B be the set of couples with incomes $100,000 -$149,999, and T be the set of couples with both employed, with incomes equal to $100,000 and over.

The answer to the first question is $P(T)$. The answer to the second question is $P(A|T)$. This is because having two incomes means both spouses are employed. We must not allow ourselves to get confused because of a change in wording. Let us evaluate these two quantities. The probability of T is equal to the probability of T given A plus the probability of T given B, as the events A and B are *mutually exclusive and exhaustive*.

$$P(T) = P(T \cap A) + P(T \cap B), \text{ as } P(A) + P(B) = 1.$$

$$P(T) = P(T \mid A) \cdot P(A) + P(T \mid B) \cdot P(B)$$

$$= .04 \times .65 + .10 \times .73$$

$$= 0.099$$

To answer our second question, we use Bayes' theorem:

$$P(A \mid T) = \frac{P(T \mid A) \cdot P(A)}{P(T)}$$

$$= \frac{P(T \mid A) \cdot P(A)}{P(T \mid A) \cdot P(A) + P(T \mid B) \cdot P(B)}$$

$$= \frac{0.04 \times 0.65}{0.099}$$

$$= 0.263$$

Of course, we can have a lot more facts, with numbers in many other income groups. Two groups makes the point simply.

10.6.2   Example of conditional probability

A doctor gives a patient a test for a particular cancer. Before the results of the test, the only evidence the doctor has to go on is that 1 woman in 1000 has this cancer. Prior(cancer) = 0.001.

Experience has shown that, in 99% of the cases in which this cancer is present, the test is positive. In 95% of the cases in which this cancer is not present, the test is negative. This means that 5% of cancer-free people test positive.

Let us rephrase this confusing statement. Let us look at the positive results. If cancer is present, 99% of the tests are positive. If cancer is not present, 5% of the tests are positive. This means very many healthy people will test positive for the cancer.

Should a person consent to taking this test? Let us look at the probabilities.

If the test turns out to be positive, what probability should the doctor assign to the event that cancer is present?

Let us look at 100,000 people, for *thinking about people is easier than thinking about percentages.* Let us look at the set of people who tested positive. We want the cardinality (number of people) of this set. Let us call this set P. We want $n(P)$, the cardinality of this set. Let the set of people who have this cancer be C. We want to find
$$\frac{n(C)}{n(P)}.$$
This is the probability of having this cancer given the fact that the test was positive. This is conditional probability. The actual probability has the denominator the size of the sample, which is 100,000. We want to calculate this conditional probability.

100 people out of the 100,000 have this cancer. Out of the 100 people who have the cancer, 99 test positive. This means $n(C) = 99$.

Out of the 99,900 cancer-free people, 5%, or 4995, tested positive. Again, 4995 people who are cancer free tested positive.

The sample size, the number of positive, is 99 + 4995, or 5094. That is, $n(P) = 5094$.

This probability is $\dfrac{n(C)}{n(P)} = \dfrac{99}{5094} = 0.0194$, or 1.94%.

This means out of 100 people tested positive, almost 2 will be have the cancer. In other words, the conditional probability that you have the disease if you test positive is 1.9%.

We see now that the probability of cancer given a positive test has only increased from 0.001 to 0.019. While this is 19-fold increase, the probability that the patient has the cancer is still small. Stated in another way, among the positive results, 98.1% are false positives, and 1.9 % are cancers. Question: Does this justify taking the test?

A lot of confusion exists with the probability of having cancer after testing positive. The problem was given a positive result, what is the probability of having cancer? The confusion is thinking that a positive result means having cancer. What a positive result means is that the probability of cancer increased from 0.001 to 0.0194.

Probability is a fraction. The numerator is the number of people with the cancer *and* who tested positive. The denominator is the number of people who tested positive (most of which are cancer free). This is conditional probability as the denominator is the number of people who tested positive, not the total number of people in the sample.

Let us change these numbers to insure you understand it. Let us replace "In 95% of the cases in which this cancer is not present, the test is negative," with 99%. What are the conclusions now? The number of people with cancer, $n(C)$ is still 99. However, $n(P)$ is different. Out of the 99,900 cancer-free people, 1%, or 999, tested positive. The sample size, the number of positive, is 99 + 999 or 1098. The probability is 99/1098, or 9.02%. This is much larger. If you have a choice between the two tests, one with negative results 95% of the time or 99%, which test would you take? What if the better test costs much more?

It is important to study the 99% example also. Students often have difficulty with this. One needs several examples to fully understand confusing ideas.

10.6.3 Another example of conditional probability

In Life Tables, one finds that in a population of 100,000 females, 89.835% can expect to live to age 60, while 57.062% can expect to live to age 80. Given that a woman is 60, what is the probability that she lives to age 80?

This is an example of a conditional probability. In this case, the original sample space can be thought of as a set of 100,000 females. Let E be the subset of the sample space consisting of all women who live at least 60 years, and F the subset of all women who live at least 80 years. Rewriting for clarity:

E = subset of women living to 60.

F = subset of women living to 80.

Note that F is a subset of E:

$F \subseteq E$.

Putting in the numbers, we have

$n(E) = 89,835$

$n(F) = 57,062$

$$P(F \mid E) = \frac{57,062}{89,835} = 0.6352.$$

Thus, a woman who is 60 has a 63.52% chance of living to age 80. Of course, if she made it to 60, she has a greater chance of making it to 80 than a young woman's chances.

10.6.4 A study involving dementia

Here is an article giving results of a study various brain diseases. I discussed this with my mathematics students. I found that the conclusions by correctly applying the mathematics were very interesting. For the purpose of this discussion, I am not interested in the research details, but merely trying to read and understand the numbers. The public incorrectly understands the data, and is unnecessarily troubled.

A Population-Based Study of Dementia in 85-Year-Olds
Ingmar Skoog, Lars Nilsson, Bo Palmertz, Lars-Arne Andreasson, and Alvar Svanborg N Engl J Med 1993; 328:153-158

The aim of this study was to investigate the causes, severity, and prevalence of dementia in a representative sample of 494 85-year-olds living in Gothenburg, Sweden.

Results
The prevalence of dementia was 29.8 percent (147 subjects). The condition was mild in 8.3 percent, moderate in 10.3 percent, and severe in 11.1 percent. There were no significant sex-related differences in prevalence or severity. Of the subjects with dementia, 43.5 percent had Alzheimer's disease, 46.9 percent had vascular dementia (multi-infarct dementia in 34.6 percent, dementia related to cerebral hypoperfusion in 4.1 percent, and mixed dementia in 8.2 percent), and 9.5 percent had dementia due to other causes. The three-year mortality rate was 23.1 percent in the subjects without dementia, 42.2 percent in the patients with Alzheimer's disease, and 66.7 percent in the patients with vascular dementia. Infarcts detected by CT scanning were significantly more common in the subjects with dementia than in those without it (27.9 percent vs. 12.6 percent).

Conclusions
Dementia was present in nearly a third of unselected 85-year-olds in Sweden. Almost half these subjects appeared to have vascular dementia, which may currently be more amenable to prevention or treatment than Alzheimer's disease.

Define sets:

D = {those with dementia}.

A = {those with Alzheimer's disease}

$P(D) = 147/494 = 30\%$. This is the first sentence in the results paragraph, "The prevalence of dementia was 29.8 percent (147 subjects)," where I rounded it to 30%.

We can write the second sentence, "Of the subjects with dementia, 43.5 percent had Alzheimer's disease,"

as $P(A|D) = 43.5\%$.

What is $P(A)$, the probability of Alzheimer's?
Given 147 people, and 43% have Alzheimer's, $147*.435 = 63.945$, or 64 people had Alzheimer's. The sample size was 494. $P(A) = 64/494 = 13\%$.
It means out of 100 (85 years old) people, 13 have Alzheimer's.

There are two major conclusions that the authors failed to write in the conclusions.

One is that $P(A|D) = 43.5\%$. This is the conditional probability, the percentage of people who have Alzheimer's given that the person has dementia. We see that the majority, 56.5%, of people with dementia do not have Alzheimer's. It means that if an old person is forgetful, he should not go crazy thinking he has Alzheimer's, a fatal disease, and is going to die.

The second conclusion is that $P(A) = 13\%$. It means that very few people get Alzheimer's. We read about how the elderly get Alzheimer's, making people unnecessarily frightened.

10.7   Binomial probability
We use the multiplication rule for binomial probability.

Let p be the probability of success, and q the probability of failure, where p and q are independent. Since they are independent, we can use the multiplication rule.
Probability of x successes out of n trials, if the first x trials are successes, is

$$p^x q^{n-x}.$$

This uses the multiplication rule, as the probabilities are independent.
Note that x is a whole number.
We are saying that x trials are successes and $n - x$ are failures.

The word "and" means multiply. We want successes and failures, and so we multiply p with q.

If we just care for x successes, but do not care if the x successes are first or wherever, then we multiply by the number of ways of choosing x items out of n where order does not matter. This is $_nC_x$. This gives the **binomial probability formula:**

$$P(x) = \ _nC_x p^x q^{n-x}.$$

Find the probability of getting one head in two tosses of a fair coin. The sample space is {HH,HT,TH, TT}. Let us do this two ways. One. First find the probability with head first. This is ¼. Then find the probability with head last. This is also ¼. Since we do not care which is first, we add the probabilities, and get ½.

The other way is to use the binomial probability formula. $n = 2, x = 1$.

$$P(1) = \ _2C_1 \left(\frac{1}{2}\right)^1 \left(\frac{1}{2}\right)^1 = \frac{1}{2}.$$

10.7.1 Example of binomial probability

Find the probability of families with 5 children having 4 girls.

Using the binomial probability formula, we have

$$P = \ _5C_4 \left(\frac{1}{2}\right)^4 \left(\frac{1}{2}\right)^1 = \frac{5}{32}.$$

This means that out of 32 families with 5 children, 5 will have 4 girls.

We will write down the choices.

Note that this question is different from the following: Pick a child out of a family with 4 girls and one boy. What is the probability you will pick a girl? The answer is 4/5.

ggggg
ggggb 1
gggbg 2
gggbb

ggbgg 3
ggbgb
ggbbg
ggbbb

gbggg 4
gbggb
gbgbg
gbgbb
bgggg 5
bgggb
bggbg
bggbb
gbbgg
gbbgb
gbbbg
gbbbb

bbggg
bbggb
bbgbg
bbgbb
bgbgg
bgbgb
bgbbg
bgbbb
bbbgg
bbbgb
bbbbg
bbbbb

It is initeresting to compare this result, 5/32, with the example in the book on page 624. They wrote a list of 50 random numbers of 5 digits. Let this represent families with 5 children, and associate odd digits with boys and even digits with girls. It turns out that there are 9 such numbers. This means the estimated empirical probability of 1 boy is 9/50 = 0.18. This is close to the theoretical value 5/32 = 0.156.

Here is an exercise for students familiar with computer programming. Write a program that will generate 1000 random numbers with 5 digits, and count the numbers that have one and only one odd digit. Compare the result with the theoretical result of 156.

Here is how to write the program. Use the random number generator to generate 5 random digits between 0 and 9. If there is one and only one odd digit, increment a counter. Loop 1000 times and examine the counter.

10.7.2 Another example

A fair die is rolled 3 times. Find the probability of getting a 4 exactly two times.

Let us analyze this.

If we roll a die one time, the probability of getting a 4 is $\frac{1}{6}$.

If we roll it twice, the probability of getting a 4 on each roll is $\frac{1}{6} \cdot \frac{1}{6}$. This is because of the probabilities are independent, and we multiply.

If we roll it three times, and ask what is the probability of the first two rolls being a 4 and the third not a 4, the answer is $\frac{1}{6} \cdot \frac{1}{6} \cdot \frac{5}{6}$. This is because there are 5 ways of not getting a 4.

However, to answer our original question, we want to add the probability that the first roll is not a 4 or the probability that the second roll is not a 4. This is the OR rule. We could have the not 4 on the first roll, or on the second roll, or on the third roll. We want to calculate the number of ways of getting a 4 two times out of three rolls. The OR rule is to add.

This is the number of combinations.

The answer is $_3C_2 \left(\frac{1}{6}\right)^2 \left(\frac{5}{6}\right)$.

This is the binomial probability, with $n = 3$, $x = 2$, $p = \frac{1}{6}$, $q = \frac{5}{6}$.

Note that there is an easy way, using binomial probability, and a hard way, using the various probability rules. Students that are not familiar with the easy way tend to do it the hard way.

This is similar to solving a geometry problem starting with the postulates rather than from a theorem that you should know. Some things are too hard to do just starting from the very basic things, but need to use proven theorems.

10.7.3 Tossing a coin

Let us toss a fair coin 4 times and ask what is the probability of getting 2 heads.
Let us look at the sample space S.

$S =$ {HHHH, HHHT, HHTH, HHTT, HTHH, HTHT, HTTH, HTTT,

THHH, THHT, THTH, THTT, TTHH, TTHT, TTTH, TTTT}

The explanation of each element is this. The first element, hhhh, is a toss with 4 heads. The second element, hhht, is a toss with the head the last toss. The cardinality of the sample space is $n(S) = 16$.

The event space E is

$E =$ {HHTT, HTHT, HTTH, THHT, THTH, TTHH}

We count the elements of E to find cardinality of the event space. We get $n(E) = 6$.

The probability is then

$$P(E) = \frac{n(E)}{n(S)} = \frac{6}{16} = \frac{3}{8}.$$

Note that this is not ½ as one may guess!

Another way of looking at this is noting that we can use the binomial probability formula. We have 4 trials, as we toss the coin 4 times. We want two successes (two heads). This is the binomial formula where $n = 4$ and $r = 2$. The answer is $\frac{3}{8}$.

In this example, the binomial formula is easier than explicitly writing down the sample and event spaces.

Note that the sample space has 32 heads. Theoretically, if we toss a coin 64 times, which we did in order to generate our sample space, we expect 32 heads. The probability $P(E)$ discussed above defines the number of events with two heads, not the number of heads divided by the total number of tosses. This is why the probability is 3/8, not ½. We must be careful what questions we ask, as different questions have different answers.

10.7.4 z-score

When we discuss binomial probability $P(x)$, where x is some quantity, such as the length of a pipe, we may find it easier to talk about $P(z)$, where z is a dimensionless variable: $z = \dfrac{x - \mu}{\sigma}$.

Here μ is the mean, and σ is the standard deviation. The variable z tells us how far the item is from the mean, but instead of feet, we measure in standard deviations.

For example, given pipes with the average length of 5 feet, and the standard deviation of 0.02 feet. We can write the average length of a pipe as 5±0.02 feet. We can speak about the probability $p(5.02$ feet$)$, or we can talk about $p(z=1)$. That is, we can talk about the probability of finding a pipe with length 5.02 feet, or length with z value 1.

10.8 Statistics

Students must be told that statistics is very different than other subjects. It is not possible to take simple examples that you can calculate with a pencil. They have to spend time at home thinking about the stuff.

All knowledge is based upon rational thinking, based upon rational principles. This is mathematics. A mathematical system must be consistent. Find one small inconsistency, and it is invalid.

To verify the truth of a mathematical system, we perform experiments and observations. Now we find only partial agreement, never complete agreement. Note that the math must be 100% consistent, yet the science is never in full agreement with observations and experiments. We compare the observations and experiments with the math, using statistics, to determine our confidence in accepting the science. E.g., the mathematics predicts the result is 5. The experimental results are 5.001, 4.99, 5.002. We say the errors are small and accept them. What is meant by small? This is the subject of statistics, which is not this course.

All knowledge is based upon statistics. There is no absolute truth, only statistics.

10.8.1 Regression

To understand one of the key ideas of statistics, consider the following.

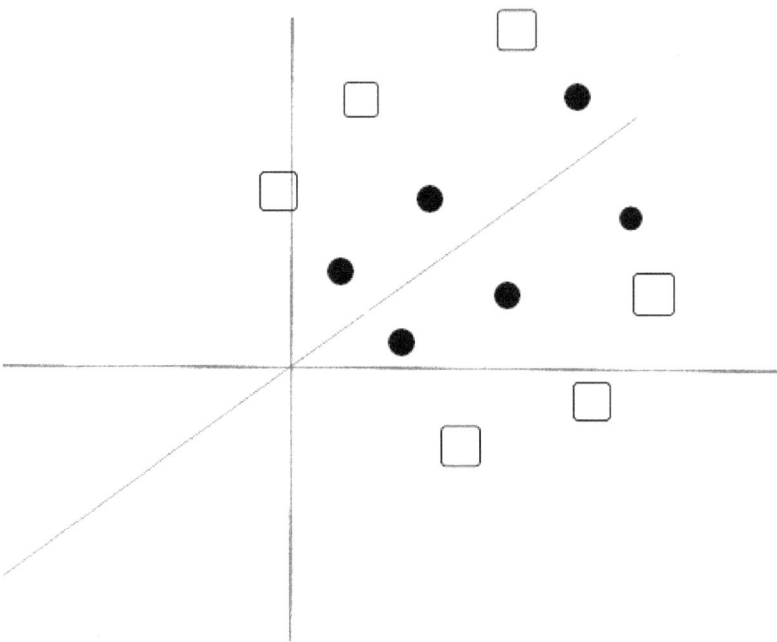

Here we have a coordinate system, x-y axis. We draw small dots. Let us look at these dots. The diagonal line is the best-fit line that can be drawn between these dots. Some dots are above the line and some below, and the average distance is zero. When we draw this line, we can predict the future. We go somewhere on the x-axis, and get the probable value of the dot on the y-axis.

This line is called the *regression* line. As you recall, if we specify two points there is only one line going through these two points. In other words, two points define a line. If we specify many points, there is no line that will go through all these points. One solution is to choose two points out of these several points, and draw the line between these two points. This is what we have done above in the chapter on lines. We looked at restaurant sales during one year, and sales during a recent year, and drew the line. We used this line to guess sales during the current and future years. This is easy. The drawback is the rejection of the information from the many other years. Sales went up and down over the years, and we wish to predict sales.

The mathematics of *regression analysis* permits us to do this. This complicated mathematics is very easy with the use of calculators. Input the many points and the calculator gives the line, written in the usual form

$y = ax + b.$

The parameter a is the slope, which we wrote in chapter 6 above as m. The parameter b is the y-intercept, the location where x is zero. These two parameters, a and b, are called the regression coefficients.

Now let us look at the small squares. The diagonal line is also the best-fit line. Yet there is something different about the small squares. They are spread out further. The way we handle this is to take the average of the squares of the distances (and then take the square root). This way lines above and below contribute, and do not cancel (squares of numbers are always positive). This is called the *standard deviation*. The standard deviation of the small squares is larger than the standard deviation of the small dots. It also means that when we guess the future of the small squares, we will have a larger error.

10.8.1.1 Standard Deviation

To understand standard deviation, let us look at something simpler. Instead of a list of points (ordered pairs of numbers), let us look at a list of numbers, and ask what is the average. Let us look at a very simple list, consisting of three numbers. We like to explain mathematics by looking at the simplest thing we can, and a list consisting of three numbers is the simplest thing.

Suppose there were 3 students in a class, and all got a grade of 70 on a test. The average grade is the sum divided by the number. The sum is 210. We divide by 3 to get the average 70. This you have learned in middle school. Let us write this using mathematical notation.

Let x_i be the grade of student i.

The average grade, $\bar{x}$, is $\bar{x} = \frac{1}{n}\sum x_i$,

where there are n students in the class.

Σ is the Greek capital sigma, and it means sum.

In another class with 3 students the grades were 70, 60, 80. We take the average, and again find the average grade was 70.

However, in spite of the fact that both classes had the same average, there is a big difference between these two classes. In one class, all the students got the same grade. In the other, there were large differences between the students. We would like to find a mathematical way to show this difference.

Here is how we can find this. First we rewrite the average formula:

$$\frac{1}{n}\sum_{i=1}^{n}\left(x_i - \bar{x}\right) = 0.$$

The sum of each grade minus the average is zero, as sum got above the average and some below. This sum contains positive and negative numbers, and that is why the sum is zero. Suppose we sum the squares (and then take the square root to undo the squares). Now we will have only positive numbers. We can prove this formula using calculus:

$$\sqrt{\frac{1}{n}\sum_{i=1}^{n}\left(x_i - \overline{x}\right)^2}$$

In our case,

$n = 3$, $x_1 = 70$, $x_2 = 60$, and $x_3 = 80$.

Using Excel with STDEVP we get 8.2, rounded to the nearest tenth. STDEVP means Standard Deviation. P means population, the entire class. Without the P, that is, STDEV, it means we look at samples.

This means that the average grade is 70 ± 8.2. That is, the average grade is 70, and the standard deviation is 8.2.

The advantage of taking our simple example is that we can calculate the result without using a computer.

The result is the same as the Excel result:

$$\sqrt{\frac{200}{3}} = 10\sqrt{\frac{2}{3}}.$$

In the old days a few decades ago prior to computers, this calculation was a formidable task. Now it is very easy. Just enter the numbers in a spreadsheet like Excel, press a few buttons, and you get the result immediately. It is also easy to do statistics with programmable calculators. What is important is that we understand what is going on. Just pressing buttons can easily give nonsense.

Here is an example of general misunderstanding. A survey that the percentage of people supporting position A is $40\% \pm 5\%$, and the number of people supporting position B is $42\% \pm 4\%$. It is incorrect to say more support position B, for 42 out of 100 people support position *B*, and only 40 out of 100 support position *A*. Some people use the phrase "statistical dead heat." It would be better to say that both positions are mathematically equal. That is, using the mathematics of statistics, both are equal.

Now that you understand this simple example with three numbers, you can appreciate that instead of 3 numbers we can have 30 or 250 numbers. You can also understand that instead of simple numbers, we can have a large number of points. Instead of an average with a standard deviation, we now get a regression line. We have mathematical numbers called regression and correlation coefficients telling how good the points make a line.

Textbooks tend to start discussing statistics starting with talking how to present large amounts of data, such as various visual displays of data. My understanding of mathematics is to start with the very simple, and to generalize and abstract to move to the more complicated things, not starting with the complicated.

10.8.1.2 Some web sites

This applet demonstrates least squares regression.
http://hadm.sph.sc.edu/courses/J716/demos/leastsquares/
leastsquaresdemo.html

A good mathematics explanation:
http://mathworld.wolfram.com/LeastSquaresFitting.html

An example:
http://www.physics.csbsju.edu/stats/least_squares.html

Index